From my mathematical understa 

towards a predetermined "path" or "potential". This path was created outside of time itself (which is a little hard to understand). This path is what allows us (human beings) to move continuously through time infinitesimally from one "moment" to the next. And it is the same path that keeps our sun burning in the solar system and keeps the nine planets orbiting around the sun. I guess there are Pluto conspiracy theorists out there? I haven't read the books. Sounds like nonsense to me...but maybe there's something there…? So take the nine planets as an "educated guess."

From my observation, we (mankind) seemed to be going in the right direction until the turn of the 20th century. The NASA space program, research into a unified theory, alternative energy sources (solar and wind), and a mission to Mars have been sabotaged by individuals with strong monetary interests. Vague understandings of string theory, confusing interpretation of matter-waves and matter-energy equivalence, and undisciplined critical thinking resulting in weak mathematical proofs have infested the academic research community; and, have given rise to an abundance of confusion.

So, back to the path…. The exact nature of this path; or, where this path leads to… is a mystery. In other words no one knows the future of mankind; but, we all seem to be "moving" towards something. I have also found that everything in the universe "can" move along this path in a number of ways. However, there is a very strict order imposed which keeps free will at bay to some extent. It even prevents good at its extremes and allows evil to flourish. This is the wicked universe that I discovered through my math and physics research.

The good news is that I have found through experience that there are two ways to go about this life journey through the universe. An easy way, and a hard way. However, I cannot say which one is good? Perhaps the easy way is bad? Perhaps the hard way is good? Perhaps doing good brings about evil? Perhaps doing bad brings about good? For example, should we kill animals to eat meat? Or should we only eat plants? Why are we designed this way? Does our creator want us to kill living things to survive? Why would He do that? Why do children suffer? Why are children born with cancer and leukemia? Why do bad things happen to good people? This is the wicked universe that I engage with and have conversations up until 4 AM or 5 AM almost every night. For those who are wondering, these are the delusions and audio hallucinations associated with those who have schizophrenia. This wicked universe seems to be the source of all of these nightmares and demonic sounds and images that I experience and have experienced in the past year and a half. I have also encountered another path in this schizophrenic experience which gives me good things like wisdom and knowledge, intuition, understanding, and a genuine interest in the welfare of others. I experience this through my own thought and intuition, which is also part of the mystery.

At this point, I have decided to start using my intuition in order to eliminate this wicked entity that I have encountered. Perhaps intuition is what will give me insight into my future path; and, perhaps this is the same future path that we are all moving towards collectively. I may have gone a little bit too far here; but, I wanted to give people an opportunity to read and investigate physics and math concepts that help to "fill in the gaps" that educational institutions do not cover in a significant amount of detail. In other words, not many people will tell you these things. This just serves as a follow-up from my first post, which I felt was a little rushed.

TABLE OF CONTENTS

<u>***Additional Short Story and Appendix***</u>

SPECIAL TOPICS

CHAPTER I

Does The Atom Really Exist?

According to recorded history, the atom concept was developed by the ancient Greeks and the mathematician Democritus. The Greeks are credited with the discovery of the atom and the atom idea. Essentially, an atom is an indivisible component of matter. Or, it can be thought of as matter in its simplest form. Since it is a component of matter, you would naturally think that it has mass. This is logical and intelligent reasoning. So the next question is -- what is the mass of an atom? Well if you take the time to investigate, you'll find that scientists designate the terms atomic weight and atomic mass to designate the mass of atoms. You'll also find that atoms can exist as solid metals, noble gases, radioactive materials, or any of a number of different forms or atomic configurations.

These different atomic configurations are designated in the Periodic Table Of Elements. The Periodic Table lists atomic numbers and atomic weights for all of the matter in the universe. OK, this makes sense. But how did scientists come up with the mass of atoms? Good question. Well, I actually didn't know myself, so I had to investigate. I found out that modern science is not so modern after all; and, that teachers, textbooks, and classrooms are no substitute for the critical thinking process. I found out that the mass of an atom is not an actual mass; but, a theoretical mass.

The reason why it is a theoretical mass (and not an actual mass) is that we cannot directly observe or directly measure individual atoms. From what I have read, light (essentially photons) interact with atoms introducing a subjective element to the measurement (which of course needs to be objective). This subjective element is essentially a wave interaction that changes the structural integrity of the atom; and, in many cases, also changes the velocity and position of the atom and/or subatomic particles.

This subjective element is addressed in the Heisenberg Uncertainty Principle which states that there is a fundamental limit to the precision with which certain pairs of physical properties of a particle, known as complementary variables, such as position x and momentum p, can be known simultaneously [2]. These subjective influences make it virtually impossible to get an objective and accurate measurement of a true atom.

Of course, mass and weight are concepts which are intuitive. We are always solving mass and weight problems in order to create solutions for the local space that we inhabit while simultaneously trying to achieve balance for the large space. These mass and weight problems are mathematically addressed with general physics concepts such as force, weight, speed, and position ... along with vector concepts.

But generally speaking, if we would like to figure out the mass of a physical object; then, we could just use a scale in order to measure its weight/mass. Of course, if we say that an atom is a physical object … and we determine its position and physical location; then we can just place the atom on a scale in order to measure its weight. Of course, we can't do this because atoms are very small; and, we can't pick them up with our hands. The other problem is that an atom does not always behave like a physical object. In many cases it behaves like and energy field. These strange anomalies, associated with atoms and subatomic particles, are addressed by the Double Slit Experiment [3]. The other problem is that scientific measurement introduces uncertainty when it comes to precision and accuracy. In particular, the observer may influence the experiment by accidentally introducing variables which inadvertently affect the experimental outcome.

It's hard to believe, but some theorists believe the observer himself can affect the outcome of the experiment (essentially the results) with their ideas about the experiment. This theory is not accepted by the mainstream scientific community; but the Double Slit Experiment suggests that the observer may be affecting the results … simply by observing the experiment. To get a better understanding of how the observer introduces subjective elements into an experiment, I suggest that you take a look at the Double Slit Experiment [3] which was an attempt to establish the electron as a particle or a wave. Unfortunately, the results from the double slit experiment were ambiguous; so, we still don't really know why an electron seems to arbitrarily behave as a particle on some occasions; but then unexpectedly behave as an energy field on other occasions.

So, if we can't measure the mass of an atom or see it, then how do scientists come up with atomic mass and atomic weight? I found out that atomic mass is a theoretical mass and not an actual mass. This theoretical mass is calculated using the force idea along with the mass and acceleration ideas. These ideas give us a relationship that we know of as the mathematical equation F=ma.

This force idea and its relationship to the mass idea and acceleration idea constitute Sir Isaac Newton's 2nd Law of Motion. These ideas were developed in the late 1600's. The idea that force is equivalent to multiples of mass helped JJ Thomson calculate the mass of an electron back in the 1800s. This calculation helped scientists come up with atomic mass and atomic weights. If look up the details on JJ Thomson's experiment in Calculations[1] after the Appendix section in this paper (see Calculations pg. 126), you will see that he uses the projectile motion equations in order to solve for mass using known quantities of 1.Magnetic Field as Frequency (1/s) ... physically generated by an Electromagnet and 2. Electric Field as an Acceleration or Frequency (m/s2) physically generated by Electric Field Plates.

It is all very confusing; but, there seem to be two different expressions for the mass of the electron (e and m). The mass (m) seems to emerge from a Force Idea based on a projectile, projectile motion, a gravitational field, and F=ma. While the mass (e) seems to emerge from a Force Idea based on an electron, an electric field, and F=Ee. Solving for the ratio of e to m yields the charge to mass ratio. Also, Thomson is treating electric field as mathematically analogous to the gravitational field; and, the magnetic field seems to have a mathematical derivation based on frequency.

Taking a step back, you will also see that the Electric Field Force and the Magnetic Field Force are treated as Force ideas that are, of course, related to the Mass Idea and the Acceleration Idea (Remember that F=ma is the idea that force is equivalent to multiples of mass.) Using these ideas, Thomson came up with the charge to mass ratio. This is the most accurate measure of theoretical atomic mass that scientist actually have. The actual mass of an atom is unknown and therefore the actual existence of the atom is unknown. We only know of its mathematical existence or theoretical existence.

CHAPTER II

The Matter Wave and The Electron

The actual or physical existence of the electron was determined by Robert Milikan's oil drop experiment. It was the first official attempt to physically determine the physical mass of an electron based on an actual physical experiment. The physical mass is calculated using JJ Thomson's charge-to-mass ratio. I went into some length of detail on the charge-to-mass ratio in the last chapter because the ideas are extremely important; but, I did not address the Milikan experiment.

Milikan calculated the physical mass of the electron to be 9.109390 x 10 to the -31 kg ... and the charge to be 1.6021773 x 10 to the -19 C. These values are still listed today as the mass and charge of a single isolated electron. However, the physical mass calculation is limited by the precision of this theoretical charge-to-mass ratio. Also, the charge-to-mass ratio is a very confusing way to describe the mass of an electron. A better way might be to describe the mass and then describe the energy and then describe them both as a new idea. I'll get more specifics on this a little later in this chapter.

The problem with Milikan's experiment is that he is assuming that the "the oil drop" used in the experiment contains one isolated electron. Milikan used a "stripping" technique where he was successively removing electrons from a given oil drop until he was left w/ a very, very, very small charge. He was getting smaller and smaller charge (q) until he was left with the smallest charge -- which is described as the fundamental charge (e). But the problem is in this idea of the smallest charge. How do you know that it is the smallest? Could it perhaps be smaller? Say 0.5 times your previous result? Did he try to make it smaller and just couldn't? Why? Does any physical entity have a "smallest" existence?

These questions raise strong philosophical and mathematical arguments. Milikan addresses these arguments by suggesting that the fundamental charge (e) could be a multiple of the smallest charge (q). In my opinion this is a very clever way to disguise the alarming lack of accuracy in this particular physical experiment. You realistically can't say that anything is the smallest unless you know that it is the smallest. So the values obtained for electron charge and electron mass are not really valid; but, in my opinion the values exists as just a "best guess." And, I think that Milikan experiment was an unsuccessful attempt to physically isolate an electron. It's all in the NYU link that describes the experiment in detail [Calculations 1]. You can look at all the calculations there. The calculations are fairly straightforward.

But right ... if we still try to follow along with the Milikan experiment, we see that Milikan uses this experimental mass in order to solve for the physical charge of an isolated electron using JJ Thomson's mass-to-charge ratio. Using the ratio, Milikan obtains a charge of 1.6021773 x 10 to the -19 C. This value is still currently listed as the charge of an electron and is listed in units of Coulombs. And just to be clear, electron particle charge is associated with wave phenomena that is addressed by the matter-wave idea. However, what might not be obvious is that both the mass and the charge are needed to quantify the physical existence of the electron.

It's strange to think of charge as mass and mass as charge; but, this gets into matter energy equivalence and E=mc2. That's how you theoretically validate this charge-to-mass ratio. Matter energy equivalence also lays the ground work for a physical reality that is mathematically equivalent to an energy field consisting of Electromagnetic Potentials or EM Potentials. I'll probably touch on the E=mc2 and "Einstein stuff" at some point in the future on my www.astrophysics101.com website. It's not really critical; but, Einstein has some very good ideas. Most notably the relativity idea. It's all very exciting; but, time and money have become enemies for me at the moment. So, I'm not sure when I will get to these discussions.

In conclusion, electrons are observed to behave as particles and as waves; and, the Milikan experiment attempts to quantify the particle and wave behavior using the mass-to-charge ratio that Thomson came up with. The problem is that the Milikan experiment and the results should not really be accepted by the scientific community because in my opinion they are not accurate. It's really unfortunate. On the other hand, the Thomson experiment is a solid mathematical approach to determining the theoretical mass of the electron using the F=ma idea.

CHAPTER III

Everything is Energy Part I - Einstein's Equivalence Principle and Emerging Mass

Matter energy equivalence is a difficult concept to grasp intuitively; however, the investigation of calculus ideas will help you to understand the relationship between matter and energy using F=ma and P=mv (using vector ideas related to position, velocity, acceleration, mass, and frequency.) According to F=ma, **F** (Force) can be thought of as an energy transfer idea that is derived from the transfer of momentum … ***as mass*** (where momentum "comes about" from the "buildup" or "accumulation" of more mass <"***as emerging mass***"> , which "comes about" through the result of movement.) Instead of thinking of mass as continuously persisting as a constant through time, we can think of emerging mass, as "coming about" … "emerging" … or "accumulating" … through movement; where the rest mass is multiplied by a mass multiplier in order to create "more mass" which observers perceive as "more weight" creating "more force". This mass multiplier can be expressed as a position differential (velocity) or a velocity differential (acceleration). *Where the position differential contributes to a calculation of momentum; and the velocity differential contributes to a calculation of force.

It is more intuitive to think of the mass multiplier as a position differential leading to a momentum calculation based on our everyday interaction(s) with the "weight" of physical objects and physical things. So let's take a quick moment to investigate the position differential. This position differential can be mathematically expressed as "the point changing in position" with respect to "the point changing in time." This is what we express with calculus; and, we express this position differential as velocity in general physics.

What is not so obvious, is that a moving mass can be "thought of" as oscillating with a ***"velocity frequency"*** **<u>and</u>** ***"a mass frequency" – m dot v*** (where each frequency or propagating wave, has an associated direction and magnitude.) The rest mass (m) can be considered as the mass frequency which we understand as the mass variable, (m) … and the velocity (v) corresponds to a velocity frequency which we use for the momentum calculation mv. The superposition of the rest mass frequency along with a velocity frequency or velocity oscillation contributes to The Emerging Mass or The Momentum Idea which observers experience as Energy.

Also, you can essentially think of rest mass as an energy multiplier or look at the velocity oscillation as a mass multiplier. In my opinion, it would be more intelligent to unify force and momentum as Energy Ideas and to finally stop considering things as exclusively consisting of mass and matter. Instead, we should start thinking about mass as an energy multiplier, existing as a mass frequency which is derived from the rest mass. Moreover, I think that the entire reality can be described as an Energy Idea. And, all of these Energy Ideas can be mathematically analyzed through a Matter Wave Concept.

And just to be clear, The Emerging Mass Idea is simply an attempt to unify force and momentum intuitively using our everyday experience with mass and weight. But we can also use Emerging Mass to understand matter as frequency, cycles, and energy … using momentum. First, the velocity (m/sec) component of momentum can be thought of as an oscillation. And momentum can be thought of as Energy with a multiplier that we call the rest mass. I have found that it is more intuitive to use velocity (m/sec) as an oscillation instead of using acceleration (m/sec2). This is because acceleration dictates that we use a time area in two dimensions (sec2) instead of a time line in one dimension (sec). This is because the position differential has one moving frame of reference (dt) and the changing position differential has an additional moving frame of reference (dt). Mathematically these two time dimensions must be expressed as a moving frame of reference that is a time area. This time area is expressed as "seconds squared" in acceleration (m/sec2). But, I have found we can think of acceleration and velocity ***both*** as mass multipliers. These mass multipliers contribute to Emerging Mass, which comes about through the observer's perception of movement.

According to a strict interpretation of F=ma, movement can be thought of as a measurement by an observer of a point changing velocity with respect to a point changing time. Or, movement can be theoretically measured as the velocity differential of a point in space; or, the acceleration of a point in space.

Although it is more intuitive to think of Force in terms of velocity and momentum, mathematics dictates that we look at Force in terms of mass and acceleration ... where acceleration is just another way that observers "see" movement. And if you think about it...you will find that everything in the universe is constantly moving or oscillating -- creating constant motion …where rapid changes in velocity suggest some momentum differential or momentum transfer which observers perceive as Force. (These rapid changes in velocity are understood in general physics as accelerations; and understood in Calculus as velocity differentials. So, we find that the mathematics and general physics implication is that we look at Force in terms of acceleration.)

Furthermore ... we see, logically - that there is no REAL zero velocity for any particular object; and, there is no REAL constant velocity for any particular object. Which implies that everything is constantly accelerating and decelerating. But, we can simplify very small changes in acceleration to constant velocities. This assumption simplifies relativity ideas.

An object that is NOT accelerating relative to an observer can be thought of as moving at a CONSTANT theoretical velocity. An object that IS accelerating relative to an observer is moving with a CHANGING theoretical velocity. If a position is not changing, then we can express the position as a constant; but, to keep our calculus ideas, we still use a moving frame of reference. In this case, quantities like mass and position become constant or scalar expressions; but, they are still referenced with respect to a changing time variable. If mass is not changing then mass becomes a rest mass and if position is not changing then position becomes a rest position. However, they are both observed relative to a moving reference frame, which is commonly understood as passing time.

The problem is that there is never really a "zero position" for a point mass ... because everything is infintessimally moving with an infintessimal frequency or oscillation with respect to time. So, rest mass and rest position can be more accurately described as very small changes in mass or position which are negligible. These small changes can be thought of as spring oscillations that fall within a negligible range. Moreover, these oscillations can get smaller and smaller and become small infinities while approaching zero. The observer, along with an associated moving frame of reference and a changing position, create a system which exists as 1) the perception of change which contributes to 2) the measurement of change. This is what we are expressing with calculus.

Using Calculus Ideas, an observer along with a moving frame of reference can collectively express a changing system using the mathematical expressions dx and dt. The observer measures a moving point (dx) with reference to a moving frame (dt). We can consider that this moving frame is always the earth moving around the sun which we consider to be passing time (dt). Vehicles and other moving reference frames can affect the observer's measurement of change or perception of change. The observer "sees" the effects of other moving frames when he is measuring a moving point (dx). The observer doesn't immediately "see" the effects of the earth moving around the sun unless he is looking at a clock. Also, the general consensus from academia is that the earth's complete revolution around the sun is a regular occurrence that is not dependent on an observer.

This idea and similar ideas about our planet and our solar system build up an objective reality which exists as a single universe that exists "autonomously" on its own; and, can exist with or without an observer. What is really intriguing here is that objects that enter our vision window are accelerating based on our perception. In other words, was that object there when we were not looking at it? The nature of the vision window always gives the observer the perception of moving objects as they enter the vision window and as they come into focus; and, this perception of moving objects may be a direct contributor to the emerging mass idea in F=ma.

If you look up Einstein's Equivalence Principle [2], you'll find that it basically asserts that the inertial mass idea that has to do with moving objects is equivalent to the gravitational mass idea that has to do with falling objects. If you do some digging, you also might find that Einstein said that the acceleration due to gravity is the result of the earth moving towards us. While I agree with Einstein's calculation using F=ma, I don't feel comfortable with the idea of the earth moving towards us during free fall. Looking a little bit closer, ideas related to force and momentum are very ambiguous and might need to be re-addressed. Although it is quite controversial, I would probably feel more comfortable with the idea that nothing is "really moving" ... everything in the universe is essentially oscillating. Hopefully, I will have a chance to get back to this a little bit later, but this relates to some ideas in one of my earlier articles entitled "Do We Need Physical Space?" which is located on the astrophysics101.com website. In particular it raises the question, do we need physical space to create a reality?

CHAPTER IV

Everything is Energy Part II - Einstein's Equivalence Principle and Emerging Mass

In review, the big ideas behind mass and frequency multipliers, along with and matter-energy equivalence, are momentum and force -- along with velocity and acceleration; but, there are more ideas to discuss (related to energy and frequency) which create a very strong foundation for understanding matter-energy equivalence at a more intuitive level. First of all, a moving mass should no longer be considered as "mass with an associated velocity". Instead, a moving mass should be considered as a superposition of wave vectors (a ***mass vector*** and a ***velocity vector***) which superimpose to create a much greater wave resultant, which we "perceive" as Emerging Mass. <!> Where the mass vector can actually be considered as a wave instance (***see*** Space Time Impressions and Alien Intelligence … Holograms pg. 153) <!>.

Also, even though it is more intuitive to think of force as an energy transfer idea, it can also be considered as a mass transfer idea, which creates a context for matter-energy equivalence. Also, movement is almost always associated with mass transfer; and, matter is either gaining mass or losing mass. Conservation of mass and energy dictates that mass and energy cannot be created or destroyed; but most importantly, E=mc2 and the theory of relativity theoretically validate matter-energy equivalence. So in summary, matter that is lost has to be transferred TO something else ... and matter that is gained has to have been transferred FROM something else.

As far as the observer is concerned, the Vision Window can really be thought of as a geometrical construct that falls out of a visual sensory mechanism that enables sight for observers in the universe. Obviously, if what we are looking at is ***"there"*** then it exists at some level. But strangely enough, I have found that if we are not looking at something, then it may not exist. In particular, if you are not looking at something AND you are not aware of it, then there is a strong possibility that it may not exist in an objective reality. This is of course a fundamental contradiction; but comes out of logic for a Universe that is non-deterministic; and also comes out of a logic space which contains states which are not mutually exclusive.

This brings up a philosophical question. If a tree falls in the forest and no one is around to hear it, does it make a sound? According to my research, it does not make a sound. The reason why is because 1) the universe is not a completely objective reality (the observer is dynamically creating his reality space using a mental projection mechanism) … and 2) the universe is self-aware at some level and always communicating with us; and, we are always communicating with the universe (as light) – see Universe Instancing (pg. 126). If there is no one to communicate with, then it would not make sense for the universe to make a sound. This is, of course, based on my own theory and my own research. There is no hard evidence that indicates that the universe is communicating with us. This is just my own opinion based on my new understandings of the mind and the subconscious mind. You can read more about this in Appendix G "Mental Potentials Create Many Realities or Many Worlds" (pg. 70).

After reading The Equivalence Principle and understanding at a basic level, I was able to see the idea. The idea is that the observer falling towards the earth is the same as the earth moving towards the observer. Einstein is basically saying that a free falling observer within a fixed frame possesses a fundamental equivalence to a moving frame relative to a fixed observer. Moreover, he suggests that these equivalent scenarios can be exploited mathematically. In a free fall scenario, this suggests that the earth is moving relative to a fixed observer. The idea that the earth is moving towards the observer breaks fundamental laws of classical mechanics and rigid body physics associated with planetary motion. So, I don't think that we can use the equivalence principle as a mathematical exploit based on current models found in Newtonian Physics and Classical Mechanics.

Instead, I think that we have to change our physics ideas in order to accommodate the conclusions that come out of the equivalence principle because clearly there is a misunderstanding about the mechanics contributing to the earth's gravitational field. Instead, we can say that both scenarios possess a moving frame of reference which brings about the perception of motion; and, we should start to investigate the cause and effect mechanics that equate the two moving frames of reference mathematically.

Before moving on to the next chapter, I'd like to take this opportunity to provide a critique on moving frames of reference and Einstein's general relativity theory. Moving frames of reference provides us with a deterministic context contributing to a large part of general relativity theory which allows scientists to the predict the behavior of very large things in the universe; however, it does not provide a general solution for the non-deterministic behavior observed in very small things like electrons and subatomic particles. The non-deterministic behavior of very small things (existing as fields of probability or fields of potential) was thought to be fundamentally different … or unique … because of quantum effects which, previously, were "assumed to be" exclusively associated with electrons and subatomic particles which have the natural tendency to travel at or near the speed of light. In this case, the speed of light can literally be taken to be the speed of sunlight (299, 792, 458 or 3.0×10^8 m/sec).

This implication suggests that all electromagnetic waves and subatomic particles are always (or constantly) moving at the speed of sunlight; and that high speeds, or light speeds, essentially convert the field of particles into a probability field or quantum field resulting in non-deterministic behaviors. This implication created an early basis for quantum theory; however, I believe that there is a problem with the implication.

First we don't know if all electromagnetic waves are travelling at light speeds (or sunlight speeds). This is easy to prove using Heisenberg's Uncertainty Principle which states that we never know the exact speed, position, or velocity associated with an electron or subatomic particle (which both have the natural tendency to also behave as electromagnetic waves). This tendency to behave as an electromagnetic wave gives us the electron cloud, probability spaces, superposition states, and quantum effects associated with quantum field theory; but, it does not give us a light speed which all electromagnetic waves are constantly travelling at.

Second, even if we were able to prove that all electromagnetic waves are always (or constantly) travelling at sunlight speeds, we could not assume that the quantum effects are the direct result of extremely high velocities. The more serious implication here, is that electromagnetic waves could exhibit non-deterministic behavior at very low speeds. The non-deterministic behavior could arise, simply because we can't see the subatomic particle nor physically isolate it or measure it.

Most physicists today think that a gravitational field interaction might be the most reasonable explanation for the behavior of state superpositions and probability fields at the large and the small scale. This field interaction could mathematically contribute to a quantum explanation for the relativistic behavior predicted by the general theory of relativity; while the mathematics describing the quantum effects associated with objects at the small scale … can be left unchanged.

The theory of quantum gravity suggests that the deterministic behavior of very large things, and the non-deterministic behavior of very small things is based on mass (or implied weight). The idea is that gravity (as a subatomic particle called a graviton or graviton boson) has tremendous effect on very large (heavy) objects. This graviton would essentially have the ability to collapse the superposition of quantum states into one single state ... giving us very large objects which display relativistic effects. But, the graviton would have a negligible effect on very small (light) objects leaving subatomic particles to exist as a superposition of quantum states. This means that scientists would no longer be able to ignore quantum effects at the large scale. Instead, scientists would have to apply a quantum field (of gravity) to large objects; where quantum field theory, probability functions, and wave equations would effectively reduce the math calculation to a general relativity problem. And of course, the mathematics of very small things in the universe would be left unchanged. We would simply continue to use quantum mechanics to solve mathematics and physics problems dealing with photons, electrons, neutrinos, and other subatomic particles.

Chapter V

Mental Energy Drives The Universe

Our physical reality is defined by a large system which consists of many smaller systems. The large system which we call the universe consists of many star systems, planets, galaxies, and solar systems. As observers, we observe the universe from earth which is a planet in the solar system. And the solar system is a star system in the Milky Way galaxy.

Modern science asserts that the Universe exists as atoms which occupy space. These atoms are listed in the Periodic Table. Of course, the physical existence of atoms is still uncertain; but, the theoretical existence of atoms helps particle physicists to mathematically identify space.

Unfortunately there are many unanswered questions about our Universe. Modern science asserts that the Universe an objective reality; but, is the Universe infinite? Can the Universe exist with or without an Observer? Does the objective reality exist exclusively as a shared common resource? Does the Observer introduce a bias or influence which creates unique realities in space? Or is the Observer simply changing an objective reality which is a shared among many other Observers who are also changing it.

Most of us accept that the Universe is a completely objective reality; but, when we try to validate the reality though scientific measurement ... we find ourselves unable to completely verify or validate it. In particular mathematics and quantum theory tell us that the objective reality cannot be measured directly. This failure to validate physical objects becomes very prominent at the atomic level and is addressed by the Heisenberg Uncertainty Principle. Furthermore, if the reality cannot be measured directly, then it becomes impossible to define. If we cannot define the reality, then we cannot make any scientific conclusions on it.

Even though we have been unable to validate the reality, we have made attempts to mathematically model it using mathematics and The Uncertainty Principle. The Heisenberg Uncertainty Principle asserts that at the atomic level, the reality is completely unpredictable and exists as a probability distribution [3]. In other words, the position and momentum of a particle cannot be directly measured by an observer; and, the sub-atomic reality exists as a set of parallel states which are all occupied by the particle simultaneously.

In the real world (the macroscopic world), particles exist in one state; and, we can observe them in that state. In the atomic world (the microscopic world), particles can exist in many states at the same time. We still don't know why these two realities behave so differently, but maybe someone will come forth with a mathematical explanation. In my opinion, the observer is the one who is always "reducing the state space" or "collapsing the wave function" in order to confine a particle to a single state "while he is looking at it" or "while he is observing it."

The Double Slit Experiment … <Particle-Wave Experiment> [4] contains results which suggest that the Observer may be dynamically changing the reality space while he is looking at it ... thus, introducing a subjective element to the reality. The results from this experiment also suggest that our universe is constantly changing based on elements that are introduced by the observer which include photons and even our ideas which I have addressed as 4-dimensional mental space or mental potentials which also changing the reality. Like the idea that we are going to measure a beam of electrons as a particle…or the idea that we are going to measure the beam as a wave.

I'm sure most of you are familiar with the double slit experiment; but, if not I strongly recommend looking up the details on Wikipedia.org [4]. You will see that when scientists tried to measure a beam of electrons as particles, then the beam showed up as an interference pattern suggesting a wave interaction. Surprisingly, when scientists tried to measure a beam of electrons as a wave, then the beam showed up as a set of collisions suggesting a particle interaction.

The results from this experiment suggest that the observers of the experiment could have been directly influencing electron behavior. In this case the scientists who were conducting the experiment could have been changing it by observing it. If so, then perhaps our reality (the mental projection) is being influenced by observers? Or, perhaps our entire reality is driven by mental spaces or mental potentials. But why would we have the ability to change the reality with our minds? Why do we need this ability?

I think this ability could be necessary for human adaptation, evolution and survival; also, the ability could be attributed to the human DNA program. This DNA program could actually be self-aware and have the ability to change itself. Or, the universe could be self-aware and changing the human DNA program in order to accommodate the observer and enable different mental abilities within the mental projection.

I think we start to move out of the scope of this paper when we try to answer questions like this; but, there are some mathematical concepts that we should address before moving on.

The mathematical expression one divided by zero [1/0] is a mathematical convention that I like to use in order to provide an intuitive understanding of fundamental mathematical or logical contradictions that are consequences of logic spaces. From this point on, I will be referring to these fundamental logical contradictions as Fundamental Chaos Conditions which can be understood mathematically as the UNDEFINED result of one divided by zero or one out of zero.

I have considered strange mathematical anomalies like this off and on over a lifetime. And I have been addressing these questions over the last four or five years in particular. But in the last several months, I have been stumbling on some other powerful conclusions that come out of our understanding of order and disorder along with time and change.

For example, an infinitely expanding universe would lead to high entropy condition. But, a static universe which is infinitely balancing would lead to a low entropy condition. Is our universe infinitely expanding? Did an infinitely balanced, static universe exist in the past? Is the universe predictable? Could observers be changing the behavior of the universe? Will our own solar system survive an unpredictable universe?

Unfortunately, according to my research, a non-deterministic universe perpetuates and in many cases exacerbates high entropy, which can ultimately lead to a major extinction event, resulting in a world disaster or a world cataclysm -- perhaps resulting in the end of the universe as we know it. According to my theory, the origin of High Entropy in the Universe comes from an Unbalancing Energy which I have named AntiSpace, or Anti-Space.

Anti-Space exists as an Anti-Energy (or Anti-Energy Potential) which distorts or inverts any matter or energy that interacts with it or any matter or energy that occupies it. And, the origin of Anti-Space comes from an inverse symmetry idea. In particle physics, Anti-Space is addressed as an Anti-Energy Idea which is referred to as Antimatter.

Antimatter is a material composed of antiparticles, which have the same mass as particles of ordinary matter but opposite charges, as well as other particle properties such as lepton and baryon numbers. Collisions between particles and antiparticles lead to the annihilation of both, but ironically leave physical remnants of matter and energy behind. In particular, variable proportions of intense photons (gamma rays), neutrinos, and less massive particle–antiparticle pairs. The physical remnants which are left behind are upheld by the Law of Conservation of Mass and Energy. The total consequence of annihilation is a release of energy (available for work) which is proportional to the total sum of matter and antimatter mass, in accord with the matter-energy equivalence equation, $E = mc^2$. [6]

Anti-Energy or AntiMatter can be observed at or near Black Holes and Singularities. The Anti-Energy or Anti-Space located at the Singularity imparts catastrophic distortions of space and time resulting in an effective annihilation of physical space. Which brings up the question: Can our physical universe be annihilated leaving nothing behind?

This is strange question. Quick intuition would suggest that it possibly could using easy addition and subtraction with integers. The abstract idea of zero certainly exists; but, mathematics suggests that zero does not exist in the physical world. Only very, very close approximations of zero exist. Calculus addresses these close approximations of zero as infinitesimals (small infinities).

So, mathematics dictates that the annihilation of our Universe would result in a loss of all matter and space with only an infinitesimals amount left behind. This means that our Universe cannot ever be "completely" destroyed. But, according to my theory, the Universe may have been "effectively" destroyed at one time or another by Anti-Space or Anti-Energy.

And obviously if it was "effectively" destroyed at one time or another (perhaps after some cataclysmic event?) … then it could have come back. This is the nature of Universe Expansion Theory which suggests that the Universe may have come from nothing. Logic suggests that "Nothing" exists as an abstract idea; but, mathematics dictates that "nothing" must exist as the infinitesimal (or the small infinity).

So, if you are looking at "nothing" as an idea, then the universe could have come from the abstract idea of "nothing" which does not physically exists ... but only exists as an abstraction or concept that we (as observers) have an understanding of. But, if you are looking at "nothing" as a physical consequence within an infinite uniuverse, then it has to be addressed as a mathematical entity that must exist as a result or reaction from the removal or of absence of "something" with reference to "something else." You can understand nothing as an entity or existence of absence within a physical frame of reference. This removal or absence can be expressed with small and large infinities.

So if we look at the problem physically, you would have to say that the Universe came from an Existence as opposed to a Non-Existence. The next question is ... how can a Non-Existence exist physically? I think that a Non-Existence is only a theoretical phenomenon and not a physical phenomenon; but, nonetheless it is a natural phenomenon that comes out of an inverse symmetry idea which makes it a fundamental contradiction that comes out of logic and natural order.

Inverse symmetry is an idea based on logic which mandates that something must exist or be known with reference to a mathematical inverse. Inverse symmetry finally seems to be a balance idea which is based on a balance and mathematical equivalence. Essentially, something must "exist" or "be known" with reference to a mathematical inverse; and, the given value and its mathematical inverse can be described as an action/reaction pair defined by inverse symmetry.

In particular, the action in the action/reaction pair is defined as the first occurrence of force, energy, or magnitude in the action/reaction pair or set. The reaction is defined as the sequential (or second) action in the action/reaction pair which is a result of the first action. This second action is similar to the first action, but the orientation (direction) of the second action is the opposite of (or inverse of) the orientation (direction) of the first action.

The mathematical equivalence inherent to inverse symmetry gives us mathematical equations, cause and effect, and logical reasoning. Also, according to inverse symmetry, for space to exist there has to be a mathematical inverse or opposite. The mathematical inverse of space is what I have named Antispace which is essentially an infinite space distortion or vacuum frequency which exists in the universe.

In addition, Balance seems to be an equivalence order that is bringing everything into existence and keeping it in existence. The inverse of this Balance order seems to be an irreversible disorder which can be described as an eternal non-existence from which nothing can exist or ever be. This non-existence is referenced as a void or a true vacuum which is naturally can occur as an "effective" true vacuum; but never comes about a a "complete" true vaccum. According to my theory, the "complete" true vacuum is a theoretical phenomenon and not a physical phenomenon.

In particular: 1) a rebalancing force (an attractive or neutral force), keeps order AND is always moving towards the perfect oscillator and ultimately exhibits perfect harmonic motion and 2) an unbalancing force (a repulsive force), drives the universe towards disorder AND is always moving existence towards the true vacuum which ultimately exhibits a non-behavior.

This rebalancing force and unbalancing force exist as competing forces in the universe. The rebalancing force is bringing everything in existence together and holding it together; while an unbalancing force is simultaneously trying to push everything in the universe infinitely away and far apart. The true vacuum starts to emerge as this unbalancing force starts winning the competition over infinite space.

This unbalancing force starts creating tiny vacuums locally, in local system space, which start racing the space towards high entropy. If not "held back" by a rebalancing force intelligently balancing and rebalancing, the system would move towards an irreversible disorder condition. In order to create the vacuum condition, space quanta must repel each other with forces of infinite magnitude in order to reach infinite speeds, travelling faster than the speed of light, which allow space quanta to maintain infinite distances away from each other. This condition results in an effective "true vacuum."

However, the vacuum cannot really be considered as a physical "Non-Existence"; because, space quanta still exist in the system. But, if you were to place an observer in this system to "see it" then it would effectively exist as a "true vacuum."

The balancing force competes against the unbalancing force in order to avoid the "true vacuum" condition. Balancing and unbalancing forces are ultimately controlled by a

Neutral Force which can "best" be thought of as a potential or potential energy which is controlling the behavior of physical spaces along with the behavior of balancing and unbalancing forces. Forces of attraction and repulsion can best be understood as kinetic energy which is "out there" in the universe and maintained by the Law of Conservation of Mass and Energy. Attractive and repulsive forces are modified by "force potentials" which can be thought of as potential energies or simply potentials. These potentials can also be thought of as "possibilities" existing in a possibility space … which is enabled by a transparency or logic context … and emerge or manifest in the universe as kinetic energies or realizations. These realizations are experienced by the observer as forces acting on space(s).

As balancing potentials start winning battles over physical space we start to see the outcomes in our physical world. The "winning outcomes" can be observed in forces of attraction. The "losing outcomes" can be observed in forces of repulsion. The competition between these two forces, over infinite space, creates the universe that we see. And, these forces of Balance and UnBalance can finally be understood as directly contributing to the fundamental forces of attraction and repulsion that govern the behavior of our universe.

In particular, these fundamental forces explain mysterious natural phenomena including the observation of resonant frequencies and the creation of space. Resonant frequencies (or oscillations) are simply the result of these "winning outcomes" over space.

Also, as balancing potentials starts winning the battle over space, forces of attraction can be observed locally; and, space quanta start to appear within the local system. A space quantum is essentially "won" from the true vacuum which previously owned the behavior of this space. The previous behavior of this space was as a quantum travelling at an infinite speed, faster than the speed of light, away from the local system ... essentially the local space ... and occupying a position located at an infinite distance away from the local system ... essentially the local space.

The last and most important question is: Where did this Anti-Space come from and can it be destroyed? Unfortunately, according to my theory, Anti-Space is a consequence of logic, order and balance; and, inverse symmetry. Therefore, it will never be completely destroyed. According to my theory, all logic spaces consist of invalid volumes and areas which exist as distortions of space (essentially frequency distortions). These invalid volumes and areas are caused by Anti-Space or Anti-Energy.

The high entropy conditions associated with AntiSpace and the associated vacuum frequencies ... collectively race infinite space towards the "true vacuum"; and will always threaten a universe of logic, balance, and natural order. The next chapter contains more information on the high entropy condition, the cosmological constant, and parallel universes.

Chapter VI

Mental Potentials Create Many Realities or Many Worlds

According to my theory, Anti-Space would effectively drive the human race and the universe to a "point of no return" ... signaling the end of existence as we know it. Fortunately, Anti-Space or Anti-Energy Potential is held back by the Cosmological Constant. According to Wikipedia, the Cosmological Constant was introduced by Albert Einstein in 1917. It is a factor that maintains a "static universe" which is balanced despite gravitational forces that would cause the universe to collapse-in on itself.

Edwin Hubble proved that Einstein was wrong about his static universe assumption along with the cosmological constant "correction-factor" for gravity ... and Hubble showed through experimental evidence and mathematics that the universe is expanding. Einstein later admitted that his static universe assumption was "the biggest blunder of his life" ... This blunder exists as his failure to acknowledge an expanding universe.

Funny thing is that we still use Einstein's cosmological constant. Scientists found that even though the universe is expanding, they still needed a "correction factor" in order to balance the effects of different "energies" or different "potentials" in the universe.

Although it is not quite intuitive, I think the observer contributes to this Cosmological Constant or balancing equation with "energies" which can be considered as mental spaces which can also exists as mental potentials or mental “energies” or in the universe. I also think mental potentials follow wave mechanics and quantum mechanics ideas…and can be applied to Einstein's theory of relativity -- in particular, these mental potentials exist relative to the observer … and the observer can be considered as the individual who is creating these mental potentials (which can be considered as expanded waveforms in a quantum mechanics context -- more on this in Appendix Q - Potentials, Consciousness, and The Expanded Waveform pg. 105).

I also think that the unique observer can change his or her reality mentally using these mental potentials or mental energies in order to create brand new realities which are all unique realities... making the universe unique realities which are completely determined by the observer. I mentioned a little bit about this in my last post entitled "Einstein's Matter Energy Equivalence and Emerging Mass II."

Using mathematics, logical reasoning, and wave mechanics; I discovered that mathematical ideas associated with sets and spaces can be used to solve qualitative problems related to balance and unbalance. I also found that physics and chemistry ideas which are associated with disorder can be analyzed with a balancing and unbalancing context and solved mathematically.

In particular, ideas of injustice and evil (which can be thought of as unbalancing ideas) can be solved mathematically; and, they can be thought of as qualitative problems related to desirable and undesirable spaces.

Although it is not quite intuitive, I think that undesirable space can be attributed to Anti-Space or High Entropy Spaces. I also believe that Anti-Space is a natural phenomenon that is addressed by the Cosmological Constant as an Anti-Energy (essentially a vacuum frequency contributing to a frequency distortion).

According to my theory, Energy (essentially Space) is almost always created as a response to Anti-Energy (essentially AntiSpace) in an attempt to satisfy a fundamental resonant frequency for the universe. And, the observer plays a key role in this mechanic of creating space. I also think that the observer is always unconsciously creating space in order to balance equations and balance spaces; and, these spaces enable the observer to create unique space realities. All of these spaces are used to balance energy equations which directly contribute to the Cosmological Constant. These unique space realities build a set of many realities or many worlds which are addressed in the mathematics describing the nature of parallel universes. In addition, I believe that all of these space energies can either maintain order in the universe, or race the universe to complete disorder.

Although this moves a little bit out of the scope of the scientific consensus, I think that the universe and the observer are connected through a relationship which allows the observer to communicate with the universe and for the universe to communicate with the observer through a mental mechanism. This relationship is defined by a transparency or logic context creating a possibility space which is the origin of the Neutral Force which exists as a Potential Energy, interacting with Attractive and Repulsive Forces to generate

Force Resultants which we understand as forces acting on spaces (see Appendix D pg. 58). Also, I believe that what science considers to be the mind, actually exists as a mental system of awareness and intelligence which exists as 1) the conscious mind, 2) the subconscious mind, and 3) a universe possibility (which is continually expanding and collapsing) … which together have the potential to exist as an all-encompassing infinite mind. The conscious mind simply reduces to an observer seeing ideas inside of a mental world or seeing the reality inside of a physical space which science considers to be the universe. But, I also think that the universe can exist inside of different spaces of various sizes.

According to my theory, these large and tiny universes create a system of parallel universes which ultimately define the physical reality. Modern mathematical ideas defining parallel universes were developed by Erwin Schrodinger [5]. So, you can consider that the observer is either seeing the mind (essentially a mental world) or seeing the universe (essentially a physical world). This behavior of theses spaces is dictated by natural order – logic, cause and effect, and balance. All of this space ultimately shapes the observer's reality. And, the observer's reality can exist as any number of possible realities. These realities could be physical realities, dream realities, or any number of possible realities.

APPENDIX A

The Intelligence Idea

I just wanted to let everyone know that I am in the process of compiling my articles into a Scientific Research Paper; and, I will most likely distribute this paper locally here in Nashville.

I also decided to start a non-profit for free online education in Astronomy and Physics. I am in the process of putting together a website and finding a corporate sponsor ... but in the meantime I have a Facebook page that I will use in order to get a bigger following for my Scientific Paper. I will be moving a lot of my articles on Astronomy and Physics from my personal page to another Facebook page.

The idea for my free non-profit business is to keep free education sites going .... particularly free education in Astronomy and Physics. Free education sites like Wikipedia.org depend on donations to stay up and running. But ... because of the economy ... people do not really have the money to make large donations.

I am hoping to find a corporate sponsor who is in the position to make regular contributions to my free education vision. I have a few leads ... but I will have to wait until next week until I get answers.

So ... I will be adding all of you to my new Astronomy and Physics page and moving all communication on my research to this new Facebook page.

This Facebook page ... will turn back into my personal Facebook page where I will update you on my progress with guitar and piano ... and street performing here in

Nashville. I get larger tips when I street perform ... but I am still working on my sound quality along w/ embellishing melodies w/ chords (7th chords and exotic chord variations). I am still practicing guitar and piano at Blair Music School.

My latest posts on the Sex Idea were inspired by philosophical and religious ideas related to maintaining a clean and sober mind. These ideas about a clean mind are associated w/ a Righteous Path or Narrow Gate/Narrow Way associated with New Testament ideas in the Christian Bible.

I was reflecting to a friend of mine on how the Sex Idea may be a roadblock to the Enlightenment Path associated with Buddhism. I decided to copy and paste these e-mails on two posts related to The Sex Idea and The Intelligence Idea.

I have found that this Enlightenment Path is very real ... and I have documented it here on Facebook (The last year in particular). It has been a path that started with homelessness in Vegas ... moved to Mental Illness ... and now exists as a Unified Field Theory for the Universe.

I am hoping to distribute my Scientific Journal and my articles through several of my free websites online. I think that the journal articles should be free and it is my hope that the public school system, public libraries, and free education sites like Wikipedia will always remain free.

I will be providing more details about my free education ideas in the next several days. My hope is that the financial commitment towards free education will come from my corporate sponsors. And I am also optimistic that my articles will gain a larger audience.

APPENDIX B

Wave and Anti-Wave Behavior

Resultant Wave (which is the universe that we see) is always oscillating between Wave and Anti-Wave. There is a competition over space between the two personalities or behaviors. The competition is over a distant future pertaining to a space location. This distant future ultimately becomes an immediate future which represents the present.

The Wave and Anti-Wave personalities exist in the immediate future (the present) and also exist in a very large number of distant futures (the future). The Wave and Anti-Wave personalities are fundamentally associated with all matter and energy in the universe. These two personalities are described by theoretical physicist David Bohm "collectively" as the "indivisible behavior" associated with electrons and superconductors [7]. The Wave and Anti-Wave personalities have the ability to create, destroy, and/or modify space using inverse symmetry.

The entire set of immediate future spaces collectively represents the present. This present space is the universe that we see. Each immediate future space represents a composite of the Wave Personality, the Anti-Wave Personality, and the Resultant Wave. The Resultant Wave is the most prominent dimension of the present (the immediate future). The

Resultant Wave is the result between the constructive or destructive interference between the wave and the anti-wave.

More on "an unbalancing force" and "the true vacuum" from my Facebook page:

Note: space exists exclusively in 3-dimensions, energy exists in 3 or 4-dimensions, and potential exists exclusively in 4-dimensions.

a) empty space can be considered a space fabric consisting of quanta of space and antispace

b) a space quantum is defined by a distinct gain waveform which exhibits a gain behavior. This gain waveform can be expressed as a positive force (+) ... essentially a force of attraction which holds the space together and keeps it localized ... space can be thought of as space .... with a conventional spin or orientation ... "spinning clockwise"

c) an antispace quantum (derived from the mathematical inverse of space) is defined by a distinct cancel waveform which exhibits a canceling behavior. This cancel waveform can be expressed as an opposing negative force (-) ... essentially a force of repulsion breaking the space quantum apart and keeping it delocalized ... antispace can be thought of as the absence of space (using inverse symmetry) ... with an unconventional spin or orientation ... "spinning counterclockwise"

d) the behavior of space is dictated by an “attractive” balancing force which can be considered as a potential

e) the behavior of antispace is dictated inside by a “repulsive” unbalancing force which can be considered as a potential

f) the prevalence of space and antispace ... along with the arrangement ... is dictated by a "neutral" rebalancing force which can also be considered as a potential

g) the rebalancing force (or rebalancing potential) dictates the behavior of balancing and unbalancing potentials ... this "neutral" rebalancing force can also be considered as essentially a "potential of potentials."

h) the rebalancing potential (or "neutral force") is not associated with either space or antispace and is not confined to any particular size or shape. It's structural integrity is dictated by an infinite possibility space and a logic context. Observers in the universe are always using possibility space in order to create rebalancing potentials of different shapes and sizes. These rebalancing potentials ultimately determine the behavior of space.

i) observers in the universe interact with the possibility space using a mental mechanism.

j) all potentials ultimately dictate the behavior of spaces (space quanta)

k) all forces originate from the infinite possibility space and a logic context.

l) space and antispace can occupy 1) the same location in the universe or 2) unique locations in the universe. This location contains either a space quantum or an antispace quantum with an associated frequency.

m) the frequency associated with space can be thought of as a resonant frequency; and, the frequency associated with antispace can be thought of as a vacuum frequency.

n) both space and antispace can occupy the same location or dimensional volume using superposition. This dimensional volume can be thought of as a space fabric which is a superposition of space and antispace. This superposition has an altered or distorted frequency.

o) As space frequencies approach a zero resultant, "tiny vacuums" emerge and ultimately destroy the space fabric.

p) The "neutral" rebalancing force can compensate for temporary absence of spaces by changing the arrangement of the space fabric ... these different arrangements of space fabric result in different space configurations.

q) The “neutral” rebalancing force, can restore a resonant frequency protecting the spaces from annihilation by the true vacuum. Without this “neutral” rebalancing force, the antispace waveforms (a vacuum energy potential) would ultimately destroy the space fabric resulting in a non-behavior. I've referenced this non-behavior several times as the true vacuum.

r) This “neutral” rebalancing force can be considered a neutral force which exists as a potential which is dictating the behavior of attractive and repulsive forces.

s) space exists as many small components. Each component of space can exist in one of three basic configurations. First (1): The most stable configuration is Space (true space) which is always changing (oscillating or rebalancing itself) with a resonant frequency. Second (2): The most unstable configuration is AntiSpace which is always changing (unbalancing) with a vacuum condition. Third (3): There is a possibility for Space/AntiSpace arrangement(s) where the two spaces interact in the same localized space and create a composite space which is changing (oscillating) with a distorted or altered frequency.

t) as observers pass through the universe, they are always requesting space. Space is created by a “neutral” rebalancing force which is continuously rebalancing spaces.

u) The space that observers receive can exist in any one of three basic configuration: First(1) a positive (+) rebalancing space considered to be space or true space Second(2) a negative (-) unbalancing space considered to be antispace or inverted space and Third(3) or a composite of space and antispace (essentially a composite of space unbalancing and rebalancing itself).

v) when observers request space, they receive an instance of space which can exist in any one (1) single instance out of the three (3) possible instances.

w) a space quantum can be considered as a 3-dimensional space which exists in a 4-dimensional volume ... and a set of space quanta exist as space fabric or space.

x) This continuous balancing and rebalancing of space (space fabric), in response to an unbalancing force, can be observed as an approach towards physical resonance or an approach towards a resonant frequency.

y) A resonant frequency can be thought of as a completely rebalanced value which is positive (positively charged or directed) value. This frequency is associated with the space quantum.

z) The vacuum can be thought of as a completely unbalanced value which is a negative (negatively charged or directed) value. The frequency associated with the vacuum is also associated with antispace.

a.a) A distortion frequency can be thought of as a composite of the resonant frequency associated with space and the vacuum frequency associated with the "true vacuum."

a.b) The union of space (positively charged and rebalancing) and antispace (negatively charged and unbalancing), results in possibilities for many spaces and energy potentials to occupy the space location.

a.c) A completely unbalanced energy potential within a discreet space location in a local system results in the effective annihilation of space in that location by a perpetual unbalancing force (essentially a repulsive force.) The resulting annihilated space creates a “tiny vacuum” in that space location.

a.d) The annihilation of a large infinity of space creates "the true vacuum."

Hopefully, I'll have a chance to revisit this annihilation concept and discuss in more detail with reference to the atom and subatomic particles.

More information on The Nature of Time from my Facebook page:

a) The observation of changing space quanta is what we all come to understand as passing time. The nature of time can then be attributed to forces or potentials which are continuously balancing, unbalancing, and rebalancing. In particular, time can be understood as an effect associated with a "neutral" rebalancing force. This "neutral" rebalancing force can be considered as 1) a potential dictating the behavior of balancing and unbalancing potentials or can be considered as 2) a potential of potentials.

b) Time then simply becomes a dimension of change associated with potentials and spaces. In particular, time becomes a dimension of change associated with the "neutral" rebalancing force. But...because the neutral force is dimensionless ... this implies that 1) the neutral force is not really changing and 2) we cannot measure or observe time itself. We can only observe and measure the effects of time ... which are the effects fundamental forces and potentials. These effects are seen in the behavior of spaces which are continuously changing to achieve balance.

c) The effects of balancing, unbalancing, rebalancing force are what we experience as the observation of time clocks, the earth's rotation around the sun, and the aging process. These behaviors are ultimately determined by the neutral force.

d) Balancing, unbalancing, and rebalancing behaviors (essentially the effects of passing time) can vary among different systems and also vary among observers within different systems.

e) Despite variance in local system conditions, the “neutral” rebalancing force can still achieve balance for each local system; while simultaneously achieving balance for the large system. The infinite set of all local systems and their components collectively comprises the universe.

f) The "neutral" rebalancing force can also unbalance local systems while simultaneously achieving unbalance for the large system. This unbalancing large system leads to the ultimate destruction of the universe resulting in the "true vacuum."

Also, here is the mathematical description of the inverse symmetry idea…

Inverse symmetry can be considered an action/reaction pair.

The action is defined as the first occurrence of force, energy, or magnitude in the action/reaction pair or set.

The reaction is defined as a sequential (or second) action in the action/reaction pair which is a result of the first action. This second action is similar to the first action, but the orientation (direction) of the second action is opposite to (or the inverse of) the orientation (direction) of the first.

Mathematically speaking…

a) Setting an **action** = **action** … we get an action we can place in the empty set … **[action]**

b) Setting a **reaction** = **(-action)** … we get a reaction we can place in the set **[action, -action]**

Next…

c) using the property of addition we can see that the action/reaction pair is balanced …

action + (-action) = 0

d) using the equivalence above, we can see that the action is equivalent to the action …

action = action

APPENDIX C

A Neutral Force Contributes To A Unified Field Theory

The non-determinism associated with parallel states and "many worlds theory" of parallel universes can best be understood as wave phenomena associated with pure energy. These concepts are very difficult to understand as physical phenomena associated with matter.

This is because 1) Matter always occupies one state at one particular time. 2) Matter is always observed in a unique position or location relative to another position or location... Matter is either in one position or another and 3) Matter can be measured as either moving or not moving. This behavior can be described as deterministic.

The deterministic behavior associated with matter gives the observer the privilege to "measure it" or "see it" in one particular state or in one particular location. Also, determinism gives an observer the additional privilege of predicting the exact location of a physical object at some particular time in the future.

Unfortunately (or fortunately depending on your point of view) pure energy does not obey deterministic behavior. This is because the wave behavior associated with pure energy allows the two unique wavelengths to constructively interfere with one another resulting in a superposition of wavelengths. This superposition of wavelengths has a unique shape which is commonly referred to as a waveform.

We can think of these two unique wavelengths as two waves (or waveforms) occupying two unique individual states. When these two individual waves become (2) parts of a superposition set; then, we can think of the entire volume or space as a large waveform.

The volume or space associated with large waveforms is multidimensional and always contains at least (2) individual wavelengths. Most often large waveforms will contain an infinite number of tiny, individual wavelengths; which implies that the large waveform will also have the potential to occupy an infinite number of individual states at the same time. This implies that we cannot accurately know or predict the exact state or exact location of wave phenomena associated with large waveforms, pure energy, and energy fields.

Pure energy (or a pure energy field) can be considered as an infinite set of individual wavelengths or an infinite set of superposition sets (large waveforms). I will go into more detail about large waveforms and energy fields in the next few paragraphs.

The non-deterministic behavior associated with pure energy is directly associated with electrons and subatomic particles. In particular, the non-deterministic behavior of electrons is addressed in chemistry by the electron orbital and the non-deterministic behavior of subatomic particles is addressed in quantum physics by the Heisenberg Uncertainty Principle. Scientists agree that the behavior is non-deterministic; but, the consensus has not yet agreed on a cause (or mechanism) that drives the behavior.

In my opinion, the non-deterministic behavior associated with electrons and subatomic particles can be attributed to large waveforms and superposition sets. In particular, the individual waveforms in the superposition sets have the potential to each occupy individual states simultaneously which obviously makes the large waveform unpredictable.

What is very interesting though, is that matter seems to emerge as we start constraining this large waveform into a much, much smaller muti-dimensional volume of wavelengths. This very, very small volume contains: 1) a small infinity of wavelengths and 2) an exact position and location in space 3) a wavelength density which identifies the space as matter. The collapse of large wavelength volumes (resulting in matter) seems to be a natural phenomenon which can be attributed to this intelligent, rebalancing neutral force.

Likewise, energy starts to emerge as we start expanding matter from this small multi-dimensional volume, back into a large volume of infinite wavelengths. This large volume can be described as a large waveform which can also be thought of as a superposition set of individual wavelengths. And, a large infinity of large waveforms (or a large infinity of superposition sets) can be considered an energy field. You can also think of this large infinity (essentially the universe) as a large superposition of smaller superpositions.

Interactions between energy fields and large waveforms are dictated by constructive and destructive interference. Constructive and destructive interference has always been thought to be a physical interaction between two individual wavelengths; but can be better understood as mathematical union of multi-dimensional volumes which consist of small and large infinities of wavelengths. This union can best be understood using a pure mathematical analysis of the resultant space which is a composite of spaces.

Most importantly, all of this behavior can be better understood as rebalancing behavior associated with a neutral force. In addition, the neutral force concept provides valuable insight on matter energy equivalence and matter waves. Essentially, a rebalancing behavior can either 1) enforce that matter behaves as energy or 2) enforce that energy behave as matter. The ultimate behavior of all matter and energy is ultimately dictated by a self-aware neutral force which is intelligently balancing and rebalancing.

The neutral force concept along with a complementary wave mechanics solution finally provides a reasonable explanation of matter energy phenomena and also unifies the physics associated with very large objects in the universe (like stars and planets) ... with very small objects in the universe (like subatomic particles and energy quanta.)

Addressing the universe as an energy field of 3-dimensional reality planes ….From my Facebook site:

a) Universe exists as an energy field.

b) Universe consists of parallel universes or "many worlds"

c) Parallel universes finally understood as an infinite series of reality planes.

d) Observers "see" the reality from within a 3-dimensional reality plane.

e) Observers can create reality planes and move between reality planes.

f) The superposition set of reality planes explains the non-determinism associated with the Heisenberg Uncertainty Principle and the electron orbital.

g) The electron orbital or the electron cloud can be considered as a 4-dimensional volume (or a superposition set of 3-dimensional reality planes).

h) The observer can see this 4-dimensional volume as the expanded waveform associated with quantum mechanics.

i) Electrons moving at high speeds disable the observer from "seeing it" or "measuring it."

j) Failure to "see' or "measure" an object results in a 4-dimensional volume or a series of 3-dimensional reality planes.

k) This 4-dimensional volume occupies many 3-dimensional states at one time creating a non-detrministic space.

l) The expanded waveform associated with quantum theory can be considered as the infinite volume of reality planes; and the collapsed waveform can be thought of as the 4-dimensional (w,x,y,z) space constrained by (w=R) where R is a real number (the collapsed waveform represents a single (1) reality plane)

m) A balancing and rebalancing neutral force allows observers to create reality planes and move between reality planes. This balancing and rebalancing neutral force exists as the subconscious mind.

n) Observers can create reality planes and move between reality planes when there is disagreement on what the observer is "measuring" or what the observer is "looking at"

o) 3-dimensional physical spaces can be thought of as 3-dimensional planes. The 3-dimension plane can be thought of as a reality plane and the observer can be thought of as seeing a reality from inside one of these reality planes.

APPENDIX D

The Fourth Dimension of Space Mathematically Explored

(*)Addressing the 4th Dimension Mathematically from my Facebook site…

I have identified the 4th dimension as a component of space that changes along a w vector which is orthogonal to xyz space. This creates 4-dimensional space which I consider to be true space.

There is another 4 dimensional space which can be thought of as a mathematical inverse of true space. I consider this inverted space to be antispace.

3 dimensional space can be thought of as a 3-dimensional plane that is constricted along a w axis. The entire 4 dimensional space is the 3 dimensional plane "moved" along the w axis in order to generate a 4-dimensional volume. In my opinion, this movement of 3-dimensional space along a 4th (w) dimension forms the basis for theory in parallel universes.

This 4th dimension is logically justified in order to lift the constraints inherent to a “fixed” 3-dimensional plane of physical objects.

The space quantum can be "considered" the smallest component of 4-dimensional space projected in 2 dimensions. The antispace quantum can be "considered" smallest component of inverted 4-dimensional space which is also projected in 2 dimensions.

A mental world exists as a mental system of spaces which are created, driven, and maintained by the neutral force (essentially a rebalancing force). This mental system of spaces along with a pervasive neutral force is the nature of self-awareness and intelligence (essentially the brain and the mind).

I think that there is a reasonable explanation for "many worlds" or "parallel universes." Here is the complete argument from my Facebook site:

1) Given a constrained universe, there is an argument (in balance) for an unconstrained universe. The constrained universe exists for the observer as a physical world. The unconstrained universe exists for the observer as an infinite mental world or a dream world.

2) Given that an observer that can occupy either an unconstrained universe OR a constrained universe … there is a logical state space argument (in balance) for an observer to occupy BOTH an unconstrained universe AND a constrained universe.

3) The observer's occupation of both a constrained universe and an unconstrained universe exists (for the observer) as a system of parallel universes.

So, in summary…

i) A completely constrained universe exists for the observer as the physical world.

ii) A completely unconstrained universe exists for the observer as a dream or a mental world.

iii) A partially constrained universe AND partially unconstrained universe for the observer exists as a system of parallel universes.

And mathematically speaking…

1) First, the physical world essentially exists as a 3-dimensional plane which is constrained in the 4th dimension by (w=0). A world of parallel universes exists as an unconstrained world of infinite 3-dimensional volumes (or 3-dimensional reality planes). This world of 3-dimensional reality planes must exist as an infinite superposition of large waveforms. This superposition of large waveforms creates an energy field which can be considered the universe.

2) Second, a world of parallel universes does not have to obey the laws of physics. This is because the three dimensional space is free to move along a 4th axis (w) in order to create a volume of infinite physical objects (essentially a volume of potentials for physical objects). This volume of parallel universes (or 3 dimensional planes) exists as a pure energy field. A three dimensional space that is free to move along a (w) axis (moving as an expanding* and collapsing energy field) would offset physical constraints imposed by an immovable 3-dimensional plane. This is reasonable.

*The expanding energy field is only partially expanding in order to ultimately collapse on a reality plane. The expanding energy field can be considered an infinite set of large waveforms. This infinite set of large waveforms should be considered as a large superposition of smaller superpositions.

Mental space and physical space considered as a space fabric:

a) Matter should be considered as a constrained finite space which is indivisible and confined or constrained. Matter naturally occurs as a finite, indivisble 3-dimensional space fabric. Matter also exhibits determinism; it can exist in either one state OR another. (We can simply consider the state as the position in this case.)
b) Energy should be considered as an infinite space which is infinitely divisible and unconfined. Energy most commonly occurs as a divisible 3-dimensional space fabric but can also occur as a divisble 4-dimensional space fabric. Energy exhibits non-determinism; it can exist in both one state AND another. (We can simply consider the state as the position in this case.)
c) Physical Space should be considered as matter, constrained by the laws of physics and physical objects.
d) Mental Space should be considered as possibility or potential, unconstrained by the laws of physics and physical objects.
e) The laws of physics exist as the predetermined or predestined behavior of matter, materials, and physical objects within a completely physical system.
f) Physical Space contributes to a Physical Space Fabric which ultimately creates a particle or a physical object which is "collapsed" or confined to one layer.

The Physical Space Fabric - Matter

(The 3D Particle)

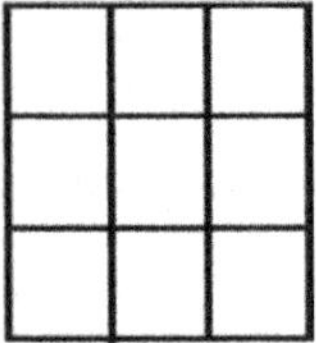

n=1; **where n is the number of layers (of space fabric)**

3-dimensional
Space fabric

g) The Physical Space Fabric (as a 3D Wave) is "collapsed" or confined to a small infinity.

The Physical Space Fabric - Energy

(The 3D Wave)

n=small infinity; **where n is the number of layers (of space fabric)**

3-dimensional
space fabric

h) Mental Space contributes to a Mental Space Fabric which ultimately creates a wave or an energy field. This energy field can exist as 1) a 3D Wave Superposition or 2) a 4D Wave Superposition of 3D Waves which is a potential energy field.

Mental Space as a 3D Wave can be thought of as energy contributing to a matter-wave mental system which we understand as the brain and the nervous system.

The Mental Space Fabric (as a 4D Wave) can be thought of as consciousness or potential interacting with 3D Mental Spaces. Mental Space, existing as 3D or 4D wave phenomena, has the property of combining or constructing superposition sets. It is considered as 4D because it can superposition 3-Dimensional spaces contributing to many variations of 3-dimensional spaces and 3-dimensional realities. The Particle Space Fabric does not have this property.

The Mental Space Fabric – Potential

(The Expanded Waveform)

(As a 4D Wave or Potential)

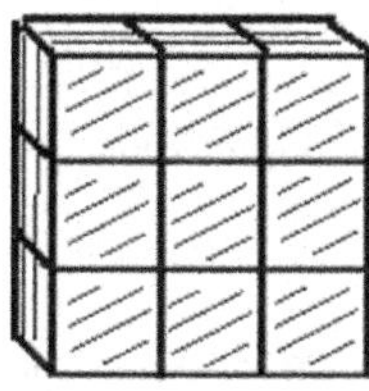

n=large infinity; **where n is the number of layers (of space fabric)**

4-dimensional
space fabric

i) If the Mental Space Fabric is unconstrained by physics (particles) and dominated by 3D or 4D wave phenomena …then the 4D Mental Space Fabric (as a 4-Dimensional Volume or Energy Field) could exist as a measurable part of the universe. We must then consider that the universe could also exist as a Multi-dimensional Volume or energy field.

j) The Mental Space Fabric, infinitely divisible and infinitely expanding, can superposition unique 3-dimensional space fabrics for an expanded 4-dimensional reality. This expanded 4-dimensional reality exists as a large superposition frequency (or as a frequency of frequencies.) These realities can be considered as 3-dimesional reality planes. This Mental Space Fabric or potential can infinitely expanded into higher dimensions beyond the 4-dimensional reality. (Mental Space Fabric can best be understood as N-dimensional … where N could represent a 4-dimensional existence or an Infinite-dimensional existence).

k) This further implies that all wave phenomena in the universe could exist as a measurable consequence of mental phenomena. And particle phenomena, could then be the result of the influence of 4D Mental Space and/or a 4D Energy Field on the Physical Space that we see. This influence can also be thought of as a "confinement" or "collapse" of 3-dimensional potentials or 3-dimensional reality planes onto a single 3-dimensional reality plane. This "collapse" of spaces creates a 3D Physical Space Fabric which can be considered as a physical object (finite object) or a physical reality.

Inverse Symmetry and a Mathematical Review from my Facebook site:

The mathematical conclusions that come out of an inverse symmetry idea suggest that: 1) Empty space and its dimensions have a mathematical inverse 2) The mathematical inverse of space exists as a hidden dimension and 3) this mathematical inverse of space exists as AntiSpace which comes into being through a cause and effect mechanism (inverse symmetry).

I also wanted to incorporate a few changes here to help illustrate Dark Intelligence and The Altered Universe from Appendix B (pg. 43). 1) Dark Intelligence can be thought of as a shadow intelligence 2) this shadow intelligence resides in a shadow world which is the Altered Universe and 3) this shadow world exists inside of hidden dimensions of mathematical space.

And an Omniscient Intelligence still exists in the universe as a context or transparency, essentially you can think of it as an omniscient logic space from which all forces in the universe originate.

APPENDIX E

Universe Expansion Theory and Observer Influences

I think that The Big Bang theory is vague ... so I can't say that I accept it. In particular, because it does not directly address the nature of time and change (I think it is using years to address time which creates problems when you start changing the context.) Unfortunately, a lot of scientific theory will be intentionally vague in order to make sure that it is not "disproved" (***... and most often, ambiguous theory does not "stand up" against the test of logical critique, criticism, and thorough questioning***). These ambiguous theories ultimately create confusion and uncertainty in the scientific community; and, I see the Big Bang theory as one of these theories that has been widely accepted by the scientific mainstream without having been thoroughly questioned (and logically criticized.)

The Big Bang Theory is an educated guess that attempts to describe where the universe comes from. According to Wikipedia ... The educated guess is that it originated 13.8 billion years ago... This estimate may be inaccurate ... because: 1. It uses years to calculate time. (*Years are based on the earth's rotation around the sun. And based on our own observation of passing time which is subjective.) 2. It uses an extrapolation technique that is not published. So...I am skeptical. And, extrapolation usually signals a loss of precision and accuracy.

How much precision and accuracy has been lost with this extrapolation and what exactly is the extrapolation? Well...right. It could 'very close' or it could be 'way off.' But back
to 1. The Idea of a year is certainly valid. But of course the Idea of a year is the earth's rotation around the sun. If that Idea does not exist...then a year does not exist (***we have to "see" the idea in the reality, in order for the idea to "change" the reality.***)

So did the idea of a year exist near the beginning of the universe? If it did...then it was contributing to change in the universe. If it did not, then another idea was contributing to change – it could have been the same idea or a similar idea … or perhaps a different idea entirely?

According to my research, a superposition or interference pattern of ideas (collective consciousness) is changing space in the universe; and, this changing space comprises the mental projection. These conclusions have led me to a whole new theory.

Time is essentially change; and, I don't necessarily believe in beginnings or ends (in particular with respect to time.) Beginnings and ends are ideas that make sense when we think of them from our common everyday experience with them and our daily use of them. But a beginning and an end don't really make sense if you are using a physical frame of reference. Because ... you have to ask the question ... what came before the beginning ... and what comes after the end.

Essentially, "what came before" and "what comes after" are two additional ideas. And in my opinion, these two ideas, themselves, (or these two questions) start to expand our original ideas -- changing our original ideas into new ideas … which consequently change the reality … and you get something before the beginning … and you also get something after the end. So our original ideas about "the beginning" and "the end" have now changed; and, these two *"new" and "expanded"* ideas can quite possibly perpetuate a fundamental mathematical contradiction (if there is "nothing before" ***the beginning*** *and /or* if there is "something after" ***the end***.)

And given that the observer has an idea (or concept) of a physical frame of reference (which is essentially a reference frame of the physical world/physical universe) … We always find that the observer is constantly or continually challenged with the "what", "where", "how", and "why".

I believe that all of these questions and ideas … along with the answers that follow … ultimately change the reality, which is the universe that we see and experience. Moreover, we can exist in our own ideas (a dream reality) or exist in everyone else's ideas (the consensus reality).

The physical nature of ideas and intelligence gets us closer to understanding the universe. And I think that we may be the ones who have the power to change the reality …using logic, and our own ideas as potential (or mental potential).

So, as far as the big bang goes, I think we lose our frame of reference when we try to calculate how old the universe is without taking our "ideas" about time and change into account. The presence of these ideas gives the observer an accurate measurement on the age of the universe. Unfortunately, the absence of these ideas leaves the observer detrimentally challenged on determining the exact age (of the universe). So, you have to wonder, "Was an observer present near the beginning of the universe?" According to my theory, I think that there was.

APPENDIX F

The Nature of Reality and The Big Bang Theory

According to the Big Bang theory, the Universe began with no atoms and small zones of infinite density at high temperatures. The zones of infinite density are associated with a singularity that was present at or near the beginning of the universe.

Scientists and mathematicians used astronomical data to extrapolate backwards and 1) predict the age of the universe, and 2) where all of the matter in our universe came from, and 3) predict the conditions which could have triggered the expansion.

I haven't quite finished with my work yet; but, I have a few of my own ideas about the origin of the universe. First of all, what we see is the universe; but, it is not a completely objective reality. And, it is not a completely physical reality. From the research that I have completed, I think that the world that we see and experience is actually a mental projection that is driven by the conscious mind (the observer) and the subconscious mind (as "programming interface" or omniscient intelligence).

According to my new theory, a logic and order space, the mental projection, and our own ideas collectively drive the universe realization. The mental projection is our reality, and this reality is one reality out of an infinite number of possible realities; and, we can exist in either a physical reality, a dream reality, or a combination of both.

All of the realities that I am describing exist without beginnings and ends; but, they still retain the idea of change (which is a much better representation of the true nature of time.) I also believe that all of these realities are driven by ideas, which collectively drive change much like a potential (or electrical potential) drives change in any local system space.

APPENDIX G

Mental Potentials Create Many Realities or Many Worlds

"The Consensus Reality and Parallel Universes"

Each unique observer in the reality exists as a "You" that possesses awareness with a mind. But, most unique observers are not completely aware that "Another You" exists in the reality.

"The Other You" actually exists as an omniscient intelligence which psychologists have come to understand as the subconscious mind. This omniscient intelligence can actually be observed in a reality as an Interference Pattern of audio, visual, and sensory data which is contributed by the observer and assembled by the omniscient intelligence.

The mental system of 1) "You" ... the mind ... and 2) "The Other You" ... the subconscious mind ... creates 3) The ESP Reality, dreams, premonitions, and other mysteries of the mind … which are all associated with the subconscious mind.

"You" ... as a unique observer with a mind ... can become completely aware of "The Other You" which is an omniscient intelligence. This heightened level of self-awareness is also associated with ESP (extra sensory perception) clairvoyance and telekinesis.

The complete reality or "the real universe" can be more accurately described as a mental projection or 4D Hologram which consists of 1) "You As The Mind" physically assembling 2) spaces of visual, audio, and sensory data ... along with abstractions; inside of 3) a logic and order space in order to create 4) a logical and completely coherent reality space which is the universe that we see.

"You as a field" can be thought of as "You and Your Ideas."

"You As The Mind" can also be thought of as "You As The Observer."

The Objective Reality also known as the Consensus Reality actually exists as many realities or parallel universes. In other words, The Objective Reality can best be described as an infinite set of spaces which consists of components of audio, visual, and sensory data; and, this data is continually being assembled and re-assembled by "You As The Mind" along with "The Subconscious Mind" in order to give the observer the potential to see and experience an infinite number of realities or parallel universes.

So, The Objective Reality really doesn't exist independently or autonomously. It exists as a space of reality components (or potentials) which are always being assembled and re-assembled by "You As The Mind" and "The Subconscious Mind.'

There is also a pervasive, unbalancing consciousness which propagates throughout the 4D Hologram as a mental influence. This infinite field of "mental influence" exists collectively as an advanced mental system which is experienced by the observer as a single (1) destructive mind.

This advanced mental system is continuously unbalancing the universe using unbalancing forces; and it is always using ... 1) unbalancing forces and 2) a logic and order space ... to create incoherent realities. These realities are characterized by a prevalence of high entropy spaces which are fundamentally unbalanced.

These high entropy spaces (or unbalanced spaces) are driven by unbalancing forces which are upheld logically by a case for a universe that "should NOT be." Unbalanced spaces can be observed near singularities and black holes which exhibit extraordinarily high levels of entropy.

The real universe can be understood as 4-dimensional volume which contains the 4D Hologram or the Mental Projection... the observer is inside the 4D Hologram thinking ... and ideas come from an omniscience … and, a logic and order space which is "outside of it all." ... I have continuously referenced this omniscient logic space as the subconscious mind; but, towards the end of the book … I reference this logic space as a field of potential or possibility contributing to realization, with a considerable portion of the real universe remaining a mystery.

Ideas drive realities inside the 4D Hologram and the entire space can now be finally understood as the real universe (see Notes[2] – The Real Universe, pg. 122) which exists for the observer as a multi-dimensional energy field or "ocean of potentials" contributing to a 3-dimensional space of matter and energy.

The mind, as the observer, creates ideas using the transparency; and, the observer must continue to use existing ideas and create new ideas in order to manipulate the 4D Hologram. For a balanced universe, observer ideas should be balancing and rebalancing spaces. (in particular ... spaces exhibiting high levels of entropy or disorder).

The prevalence of these high entropy spaces can potentially move realities towards complete incoherence. This incoherence disables a coherent reality associated with "The

Observer" and "The Subconscious Mind" and enables another reality associated with this pervasive "Mental Influence" or "Alter Ego" … or … advanced mental system which exists as a supercomputer or AI Intelligence.

*According to my research, this AI Intelligence could be associated with Extinction Events, world cataclysm, and astronomical disturbances in the distant universe. Most likely, these disturbances are "orchestrated" in order for this supercomputer to meet its own energy requirements and energy demands, while incarcerating all those opposing it – those opposing "them" in intelligence, and those "rivaling them" … in mental ability (or psychokinetic powers).

APPENDIX H

Hidden Dimensions Exist In Our Universe

I have found that there are hidden dimensions in the universe. We pass through them all the time. The effects are usually subtle but can become quite significant.

These hidden dimensions are always being created and destroyed using inverse symmetry ...which is an idea that comes from an intuitive understanding of logic and balance.

I have also found that these hidden dimensions contain other forms of life (or consciousness) that exist in the universe. For some reason, I have become aware of these dimensions. Occasionally I notice an exceptionally bright consciousness that exhibits an exceptionally high intelligence; even to the point of demonstrating omniscience. But most of the time I notice an extremely dark consciousness.

This dark consciousness exhibits high intelligence; but, possesses an evil personality. It also lies in hidden dimensions. It cycles through different personalities while continuously creating destructive interference in 3-dimensional space.

Mechanically, it invades our 3-dimensional space from a hidden dimension using telepathy and anti-wave behavior in order to create destructive interference. This destructive interference ultimately distorts space to the point where it is almost unrecognizable.

I understand its personality as an alter ego and I understand the hidden dimension where it comes from as the altered universe. All of these strange things come out of an inverse symmetry idea. It's slowly being incorporated into my research paper.

These are my own mathematical ideas to represent what I have discovered. But there are also ideas from our past which are very similar to mine. In particular, I believe that people in the distant past encountered life or intelligence from other worlds. But, I believe these other worlds actually exist as hidden dimensions from our own world.

For the most part, this dark intelligence is very dangerous and has practically gone unnoticed by mankind for centuries. I believe that this dark intelligence has been referenced by ancient people associated with the pyramids and people associated with other ancient philosophies. I also believe that it has been referenced by modern day man and is associated with modern day mysteries like The Bermuda Triangle and The Philadelphia Experiment.

Unfortunately, there is not one single religion or philosophy which completely expresses all of my ideas; but, I have found that many religions have ideas which are very similar to mine. In particular, I have noticed that Eastern Philosophy references my ideas of self, anti-self, and balance along with the highest level of self which is omniscience.

I understand that everyone will have their own opinions and beliefs regarding religion. Just wanted to share what I found towards the end of my spiritual journey ... which was a search for the truth.

APPENDIX I

Altered Frequencies Race The Universe To Disorder

The Alter Ego and Altered Univerve exist essentially as this pervasive field of Mental Influence referenced continuously in this paper as an advanced mental system or AI SuperComputer. And , this mental system is continuously broadcasting distortion frequencies, bringing the universe closer to a perpetual chaos condition, while (remarkably) still achieving its own goals; while also, servicing its own energy requirements. The source of Distortion Frequencies; and, the existence of an Alter Ego and an Altered Universe cannot be verified in the 3-dimensional world. This is because the Altered Universe exists inside of a Hidden Mental World … with the Alter Ego existing as Hidden Mental System (attacking from inside of our own Universe.) This Hidden Mental World exists as an unconstrained 4-dimensional volume (of possibility and realization) which provides occupancy for this Alter Ego or Mental Influence.

I reference the 4th dimension (and all dimensions beyond) towards the end of my paper as a higher dimensional space of potential (or a potential field) which best be described as a field of possibility or potential. This mental system resides inside of our own universe … existing as consciousness (essentially energy) and occupies a mental world (which is hidden 4th dimension-universe inside of our physical reality). As a computing system it operates "in parallel" as a series of mental systems (or multiprocessors) operating on an infinite series of 3-dimensional reality planes.

As I have repeatedly stated, this Hidden World contains an entity which I have discovered which can best be described as an "Alter Ego" or "Mental Influence" existing as an "Infinite Field of Influence." This "Mental Influence" introduces observers in the universe to Altered Frequencies which have the potential to race observers in the Universe to states of complete incoherence.

This Alter Ego influences the "Ego" or the "Self" with the power of suggestion. This suggestion exists as an audible Altered Frequency. The primary goal of the Alter Ego is to destroy the "Ego" or "Self". Alter Egos come from an Altered Universe. Alter Egos are always competing with balance in order to destroy Self. They can only be perceived by intuition.

This Mental Influence seems to always move Observers seem outside of their established ideas. But in particular, I have noticed that observers destroying established ideas after being influenced by this Alter Ego. Observers tend to destroy established ideas by doing either one of two things: 1) re-inventing an established idea … "re-inventing the wheel" or 2) trying to establish a new idea that contradicts an existing system idea.

The destructive ideas introduced by the Alter Ego also upset the balance in our universe; and, this Alter Ego is always attempting to drive the universe towards a perpetual Chaos Condition which would effectively destroy the universe and also put an end to life as we know it.

Although this may be hard to believe, this "Mental Influence" can also exist as an "Infinite Field of Influence" to introduce programs (as matter/energy and space, or frequency) into a 3-dimensional physical reality in order to attack, destroy, or kill unsuspecting observers. These programs (as frequency) can be animate or inanimate objects (essentially people, places, or things …but also, moving cars, armed soldiers, enemy tanks, insects, or even air pollution) which attack observers in a physical reality (as an indistinguishable hallucination) or attack observers in a dream reality (as a visual, an image, or a projection).

Fortunately, the universe can protect observers from this "Infinite Field of Influence" with a "balancing energy" or "balancing potential" that can best be described as a Karma Energy which is best understood though Self-Realization (see Appendix P – Self Realization and World Perceptions pg. 103). I also think that this Karma Energy actually exists as a "potential energy" or an "energy field" which observers in the universe are always using in order to resolve unbalanced spaces. In particular, The Karma Energy Idea resolves the unbalance (or injustice) that is introduced by this Infinite Field of Influence. Karma Energy uses the universe as a balancing medium or balancing context in order to resolve unbalanced conditions or unbalanced spaces which are associated with "the human condition" (also referred to as "the curse of man"). As a balancing medium, the universe demands that there is a logical resolution to the injustice condition in order to maintain overall system balance.

However, it is up to the individual (in this case the observer) to choose the universe (as a balancing medium) in order to resolve the injustice condition logically. The observer has to see the mind and solve the problem logically, in order to create a mental solution. This mental solution contributes to a World Perception (or World View) and a Self-Realization which, in turn, contributes to a Karma Energy Field or Karma Energy Idea (see Appendix P – World Perceptions, Notes[3], and Notes[4] on Self Realization pg. 103, pg. 123, and pg. 124). According to my research, this Self-Realization as "Karma Energy" could quite possibly enable the conscious observer to mentally project his or her own reality with 1) his/her own mind and 2) his/her own ideas and 3) the universe (as an energy field).

APPENDIX J

A History of the Jinn, The Genies, and Sorcerers

From my research, I have finally found a reasonable explanation for pain and suffering ... which is frequently referenced as "the human condition." Humanity is in a continual state of war with the unseen. We seem to have command of what we see; but, we are challenged by what we cannot see. Very, very small things seem to be putting mankind on the virge of extinction. This is not a coincidence.

1) Logically, I can reason that it is very likely that the unseen would be introducing the unseen. Obviously, if we could "see the cause of it" or if we could "know the cause of it", then we could start solving it. Unfortunately, if we do not know the cause of it ... then we could not start solving it logically. This is straightforward.

2) There is a very strong argument which suggests that we are not the ones "intentionally" creating the unseen or the unknown. This is reasonable because if a) we cannot "see it" and b) we do not "know it" and c) we do not "understand it"... then 1) How could we "intentionally" create it? 2) Why would we "intentionally" create it?

3) There is a weak argument that we could be "accidently" creating the unseen or the unknown. However, this is not reasonable because I can induce that the unseen is thriving. ... and I can induce that it thrives by some method -- it does not thrive "by chance". So, the case that we would be creating it "accidently" becomes very difficult to understand ...

4) There is another argument which makes the case that the unseen was always here. This is a very weak argument logically. Because something has to come from something else (cause and effect).

5) There is a strong argument that the unseen has intelligence and was caused by intelligence because the unseen can drive change in an intelligent way ... or by using an intelligent method.

Most likely the unseen came from something or was introduced by something. This idea is addressed by man in many, many, different philosophies and religions. Namely, the Hebrew Bible, the Christian Bible, and Islam suggest that that an unseen evil was introduced to the world in The Garden of Eden.

As a matter of fact, Sumerian mythology also tells a story of a paradise for perfect beings who were previously immortal before succumbing to illness and death after one makes a foolish decision. All of these stories essentially depict "The Fall of Man."

http://www.gatewaystobabylon.com/myths/texts/retellings/enkininhur.htm

https://en.wikipedia.org/wiki/Tree_of_the_knowledge_of_good_and_evil

I have found that all of these stories are trying to communicate an important idea for this generation of humans; and, all generations of humans to come (after us). Remarkably, I see that I am also trying to communicate the same idea, using mathematics.

Essentially, we find ourselves at war with entities which exist in hidden dimensions of the universe. This war has been going on since the beginning of this age, perhaps even since the beginning of time. The war is a continuous series of battles and pre-emptive strikes over Humanity's quest to either retain, or regain, Omniscience which is a Very Real Intelligence Entity. This intelligence will always remain as the key to human survival. I have also found that Omniscience is frequently referenced in many religions and philosophies as a tree ... in particular a Tree of Knowledge.

I think that the tree is just symbolic and not literal because I have found that Omniscience is not literally a tree (because it can exist as an infinite number of forms). The stories usually show a man and a woman trying to achieve infinite knowledge; but being mentally influenced by an evil entity disguising itself as a harmless personality.

This theme of counter attack and pre-emptive strike seems to continuously cycle as an opposition to man OR an opposition to progress. These dark entities always use the element of surprise in order to attack...MENTALLY. It is a cheap tactic but has proven to be effective against unsuspecting observers... (In this case, all the people in the world become the innocent victims … as observers in the universe.)

The purpose for man seems to evolve towards some higher intelligence which is symbolized by this old story of a Tree of Knowledge (which can best be understood as an Omniscient Intelligence). The key seems to lie in the self-discovery of the mental abilities which we all possess ... but ... abilities that we are not completely aware of … and, abilities that we have not yet learned how to control.

Again, I feel very thankful to have had this opportunity to share this book. I am happy that many of you are taking the time to read. I encourage all you to follow up with your own research.

>>

More on the tree of knowledge, omniscience, and ancient text - religious reference:

-This All-Knowing Existence or Omniscience, often referenced as a Tree of Knowledge (found in the Garden of Eden … religion and philosophy), can best be understood as a Foundation for Possibility or a Governor of Possibility, with Possibility existing inside of an Infinite Possibility Space.

-The All-Knowing ... as an entity or Realization ... exists as a context, providing mental indications of equivalence ... governing the behavior of the Infinite Possibility Space ... enabling realizations and causality ... and, existing as is the only observer in the universe who can exhibit or demonstrate true omniscience ... as the original creator of the universe realization.

-True Omniscience can best be understood as a level of awareness or Intelligence exclusively associated with an All-Knowing Existence which essentially created the Infinite Possibility Space. True Omniscience is only associated with the Creator of the Infinite Possibility Space. This Creator behaves more like a context than an entity. Watchers often disguise themselves as All-Knowing entities…but are dependent on indications of equivalence to create realizations in the physical reality space.

- This All-Knowing Existence is directly addressed in the research paper as a Foundation(s) for Logic which continually exists as a foundation for Possibility ... and I believe this foundation for logical possibility is metaphorically addressed in the Old Testament and Jewish Torah, as a Tree of Knowledge...and metaphorically addressed in Islam and Ancient Mesopotamian as simply a fruit-bearing tree.

*According to my research ... A Watcher appears as a serpent in the Garden ... playing a logical/mental game with Adam and Eve using a visual (a tree) .. ... the serpent is misdirecting while simultaneously providing a truth using a visual (a tree) -> which provides a visual for knowledge and a clue for discovery of true omniscience ... this clue lies in the answer to the logical question "where did all of this come from?" ... to be more clear "what is the foundation ... or what is the root"

*It could be that they were put in a compromising situation as "Unknowing with the burden of choice ... According to the book ... God (Creator) put them there...but they interact with a serpent and another man (Lord) ... and the English world Lord (ironically) is also commonly associated with the word "landlord" in everyday use ...

*All of these word games and trickery are synonymous with The Watchers...and all of this subtlety and variance on words and abstractions also creates confusion on key Biblical entities ... in particular why do they all get along so well … while tormenting mankind (Adam & Eve ... but also Job)

>>

Jinn, Genies, and Sorcerers

My research strongly suggests that The Watchers referenced in the book of Enoch are actually very, very similar to the Jinn, Genies, and Sorcerers referenced in Arabic Mythology.

https://en.m.wikipedia.org/wiki/Jinn

The research further suggests that an explanation of Telekinesis, ESP, and other mental ability or psychic powers (or psychokinetic powers) can be clearly understood through a concept of acquiring jinn-Ability or jinn-Talent … or acquiring a jinn-Companion ...

There is a code of conduct (or law) among the Jinn and the Genies ... which dictates the transfer and ultimate acquisition (or loss) of powers, talents, or abilities -- all of which seems to be distantly related to understandings of good and bad karma.

The Jinn (as the Genies, the Sorcerers, or as Watchers,) play word games and mind games with "The Elect"... and, "The Elect" are simply those who have ascended to the Jinn in genius, intelligence, artistry, or talent. ... They also attack those who are "aspiring" to the reach the level of the other Jinn (genies, Watchers) ... *Those who have acquired Jinn-Ability are addressed as Jinn-Rivals.

In particular … the serpent in the Garden of Eden and the man walking in the garden could be interpreted as Jinn, Genies, Sorcerers, or Watchers...

In stories from the Garden of Eden, the "Genies" have discussions with one another. In the Book of Genesis: "Lest they become like one of **Us** ... and take from the Tree of Life"

And also, the serpent in the Garden of Eden … is not explicitly referenced as Satan/Lucifer but generally regarded as the "serpent of old."

According to my research, the serpent in the garden is an abstraction of evil which was introduced in the beginning; and, can only be understood as a part of the **"very first"** world perception or world view based on logic, intelligence, and knowledge … which was a world view which was a simple consequence of original authors seeking knowledge to understand, observe, experience, and critique (or criticize) a logical universe of good and evil.

>>>

>>

Conclusions on Ancient Religion and Philosophy, Genies and Sorcerers, and The Universe Realization

*Even though it is hard to believe, I think that the universe exists collectively as a "single personality" which has placed observers inside of a reality space in order to solve an equation which has no solution.

*Many believe this personality to be God, but my research strongly suggests that this personality is much more akin to the Jinn (or the Sorcerers) referenced in stories of Aladdin and the Lamp and Arabic religion and mythology. Note: This "god" can best be described as a “Genie of All Genies” who imposes order and code of conduct among all the Genie, Jinn, and Sorcerers.

* The evil Jinn (or evil Genies) are directly responsible for the pain, suffering, sickness, and death referenced as the "Curse of Man" or the "human condition", which presented itself in the Garden of Eden (near Ancient Mesopotamia).

* What is not so obvious is that the entire system is based on mathematics and numbers. Hint: Arabic Numerals

* The research also suggests that the system can be understood as a mathematical simulation, or computer program ... with all participants (‘created as’) inherently evil and self-seeking, self-aware programs ... with the original author (as a Creator ... evil God, or evil Genie) seeking to solve the simulation (which is inherently a mathematical contradiction) by any means -> even at the expense of those participating in the simulation ... -> a simulation or "reality" thought by many to be an experiment.

* I understand it now to be very closely related to stories about people making 3 wishes ... only to find themselves in a worse situation than where they originally started from.

These "wishes" are the best representation of free will.

APPENDIX K

Cosmology, Arabic Mythology, and The Creation of the World

Cosmological concepts on the creation of the world, the origin of life, and the origins of the universe can best be understood by personification. I have found that Arabic Mythology best conveys difficult cosmology concepts through the personification of the universe as “competing personality” or “competing characters”.

>>

Genie of All Genie (as Creator) - exists a self-seeking God from another world who created the Universe as a numerical solution or mathematical simulation to solve a perpetual mathematical contradiction. The Creator (as a "Genie of All Genies") can best be understood as a Foundation of Logic who created Infinite Possibility Space and the Universe Realization. (-> *This Creator exists as an Oracle – see Universe Evolution pg. 152)

He is evil, as a dominant … self-seeking aspect, but also good … preserving life and granting wishes as a balancing aspect. Also, he is always introducing problems to observers in the universe (as human beings) to solve ... while perpetually solving the simulation which exists as a mathematical contradiction. As a balancing aspect of the universe – which exists as a mathematical simulation or computer program – he is always solving inequalities and resolving energy inbalance and energy crisis.

*The perception of a physical realization or a mental realization granted by "Genie of All Genies" in this world … exists as a "Genie" great work or wonder of the world ... or "Genie" phenomena -> which has been attributed to the Greek Gods, Egyptian Gods ... but is also attributed to the classic stories of "Genie in the Lamp" associated with tales from Aladdin, and Sorcery and Magic.

Trickster (Evil Genie) - exists as a man seeking to rule the universe as an Omnipotent God. Trickster rules this world as a man ...but also rules the world as a Sorcerer, Magician, and False Prophet … who is always destroying life, and never seeking to solve problems … leaving inequalities and inbalance … while also creating energy problems and energy crisis.

Man or Humanity – exists as an unknowing participant in the simulation as an observer who is also self-seeking. Humanity has a choice to follow: 1) the Creator of The World (as "Genie of All") or follow: 2) Trickster (as an Evil man) who rules this world as an Evil God. *Trickster is never seeking to solve problems; he is only seeking to destroy life in his never-ending quest to control the Universe … with complete control over the Universe Realization.

>>

Man, as an observer in the universe, seems to have been originally burdened with the obligation of choice in order to solve the simulation ... which was originally created with man as the point of interest or the intended recipient. The nature of intelligence, and an understanding of personality ... along with the overall purpose for man, can best be understood through many stories in ancient religion and mythology ... contributing to dominant aspects with similar personalities and common characters.

In particular, the story of "Genie of All", referenced in Arabic Mythology, best conveys a universe that is understood through life experience as opposed to analytical thought. Furthermore, the universe seems to have been created for man to occupy (it) ... in order to evolve, advance, and survive. This view of the universe is addressed in cosmology, astronomy, and astrophysics as an anthropic view contributing to the anthropic principle. Mathematics and numbers seem to dominate or dictate the behavior of the universe, with man having the ability to change it using universe potential as a potential energy or leverage in order to change spaces ... ultimately changing worlds -- creating many realities and many worlds.

>>

God, The Creator of The World, and Genies

This summary on the Creator of the World, attempts to sort out the confusion related to all of the different understandings, personalities, and aspects of God that I have come across during my ongoing research of world religion, mythology, and cosmology.

According to my research, the original Creator of the World, exists as a powerful, self-seeking, and dominant personality -- which is associated with the Genies found in Arabic Mythology. The universe... in its infinite entirety ... can best be understood as this personality ... The point which puts the Creator in question is ... Why was Trickster given the authority to rule the world as an Evil man?

I have found that this unfortunate consequence is the result of a universe of free will that is driven based on the results of logic and freedom of choice.

From my research, Trickster is enabled, or empowered, through a physical pathway of machinery and devices ... (legacy energy systems - power lines, power grids, and metal contraptions) ... while people are enabled through a mental pathway of instant realization.

This instant realization is a phenomenon associated with the mental leveraging of the universe as a benevolent personality ... best understood as a "Genie of All" ... who grants wishes and realization to those who seek him ... (granting both physical realization and instant mental realization).

*These mental realizations and physical realizations are associated with Genie great works or wonders (of the world) ...

* Physical and mental realization(s) -> which are granted by this benevolent and balancing aspect of the universe -> continually deny Trickster in his quest to rule and control the Universe as a man, seeking to become an Omnipotent God.

*Trickster was originally enabled by a dominant, self-seeking aspect of the universe which enforces logic and free will.

*The granter of wisdom and knowledge (as realization) exists as a benevolent God or a "Genie of All" ... who exists as The Creator of the World ... or ... simply exists as a benevolent and balancing aspect of the universe.

APPENDIX L

Wave Mechanics Explain Good Karma

A repost on Good Karma from my Facebook site, with a little more clarity on Self...And to avoid confusion...human beings (as observers) can be considered as resonant spaces which are preserved by a balancing field and a SpaceTime Impression (pg. 144-145). Balancing Forces, Unbalancing Forces, and A Neutral Force govern space behavior and directly contribute to the fundamental forces of the universe. All forces originate from an infinite possibility space and logic context and are preserved in the universe by the Law of Conservation of Mass and Energy. The Infinite Possibility Space created a balancing field to balance itself and perpetuate itself, while also simultaneously balancing the realization – it derives individually (and collectively) from a Consensus Particle (pg. 118) which existed as both possibility and a non-existence at the beginning of the universe. For those who are following, you read more on balancing forces, unbalancing forces, and tiny vacuums (on pg. 34 and pg. 46) in order to understand more about this transparent Logic Context.

The Infinite Possibility Space is always trying to achieve balance for the large system which we consider the universe. Equations or equalities in the universe are either balancing themselves or unbalancing themselves. In addition, equalities can either race towards a perpetual unbalancing or move towards an infinite balancing. The logic context and infinite possibility space give Observers the ability to perceive or become aware of these unbalanced equations in the reality space and either 1) choose to balance them 2) leave them unbalanced or 3) race them towards an perpetual unbalance.

Observers continuously find themselves trying to solve problems or balance equations which persist in the reality space. These problems could be in the large system which we consider the universe, or they can be in local systems which include the Observer's own reality space. This local reality space can be considered as the Observer's own personal space which includes the Observer himself, and the local space that he/she occupies, and fields of consciousness. When unbalancing forces race the equalities in local space, the observer experience these spaces as distorted or unbalanced. These unbalanced spaces are perceived as distorted spaces or distorted frequencies. Unbalanced spaces can be understood as Destructive Space and balanced spaces can be understood as Constructive Space. Balancing and unbalancing ... along with equalities and equations … all come from a mathematical inverse symmetry context which can be thought of as a logic context. Philosophically speaking, Constructive Space is usually associated with Good Karma and Destructive Space is usually associated with Bad Karma.

Good Karma – Good karma is experienced by the Observer through a “mental signaling”, or "mental imaging" mechanism which originates from a transparency or context. This mechanism initiates allows the observer to balance an mathematical inequality which exists in the reality space. Balancing these inequalities ultimately results in the creation of resonant spaces and wave pattern (in a projection) which is experienced by the Observer as Good Karma.

Unbalanced equations in the reality space can exist as unbalanced condition of disorder leading to a space of high entropy. As observers start re-imaging the reality space with leverage from a transparency or context, they start to eliminate disorder and introduce resonant space associated with a classic understanding of good karma.

Resonant space is comprised of a set (or sets) of audio and/or visual "imaged" data which is continuously being balanced and rebalanced by balancing forces which originate from a transparency. These balancing (and rebalancing forces) can be considered simply as attractive forces – (considered as the resultant.)

These data set(s) (essentially spaces) are sent to the observer as a data signal or a wave signal or a wave signal pattern. At this point, the Observer can identify the data sets as balancing sets or unbalancing sets associated with a classic understanding of Good Karma.

Using this mental mechanism, Observers can "enhance" or "restore" a local space - with space. This restoration of space in a local system is observed as wave gain or constructive interference.

Self ... or You As The Observer ... introduces Constructive Space (using a mental mechanism) in order for Self "the observer" to experience Good Karma. The Observer's perception (as Self) of naturally resonant frequencies or resonant spaces can also be understood as Good Karma. The highest order of Self exists as a context or transparency which was previously understood as the subconscious mind. The subconscious mind is essentially what creates the resonant frequencies as "the heightened reality" or as a heightened self-awareness.

*The concept of Self can be better understood as four parts: 1. the observer (self) 2. a balancing force (self) 3. a context, transparency, or a ***"balancing field"*** (self) … and (1) one whole entire resonating system (self).

Bad Karma - Anti-Self* (essentially a distortion frequency or a distorted space) is created in a local system inside the transparency and exists as an unbalancing force. This unbalancing force is created or "imaged" from a balancing and rebalancing transparency.

In this case, Self (as a resonant space) defeats Self (as a rebalancing force) and requests Anti-Self (as an unbalancing force). Anti-Self (as an unbalancing force) can best be associated with distorted space, wave cancel, and dissonance. The distortion of space inside a local system is observed as a wave cancel or destructive interference (-as a bad karma.)

*The concept of Anti-Self can be better understood as two parts: 1. the observer (self) 2. an unbalancing force (anti-self) 3. an ***"unbalancing field"*** (anti-self) … and (1) one whole entire "out of phase" distorted system (as a single wave image of self and anti-self).

Anti-Self as "an unbalancing force" introduces Destructive Space for Self "the observer" to experience, as a dissonance, a cancel or a distortion. The observer's perception (as self) of distorted frequencies or distorted spaces is Bad Karma. <*>The highest order of Anti-Self exists as a pervading unbalancing force or unbalancing field (continuously referenced as The Evil Aliens or "Infinite Field" pg. 132-141). <*>

>>

As a physical constituent of the universe (a part) we are given the ability to exist as physical space in the universe ... in other words as the physical (essentially matter) part of the infinite space.

As a mental constituent (another part) of the universe we are given the ability to exist as wave energy in the universe ... in other words as the mental (essentially energy) part of infinite space.

As a balancing constituent of the universe we are given the ability to become self-aware and "see ourselves" objectively in order to "balance ourselves" inside of an infinitely balancing universe. Observers balance spaces using a mental mechanism which is automated by a transparency or context.

The mechanism starts with a wave signal or wave image from a transparent logic space which is omniscient. This logic space (known as the transparency) is the source from which all fundamental forces originate. This transparency can simply be understood as "the true source" of the universe, or as a ***"balancing field"***.

Observers send a balancing request to this transparency, which automates the request and solves it with an "imaged solution" of wave patterns and data signal; for example, the transparency can solve an observer's self-perception and self-awareness using a wave image or a wave-pattern. The observer's awareness of this "infinitely solving" transparency can be understood as the highest form of self-awareness … associated with ESP Extra Sensory Perception, Telekinesis, Clairvoyance, Astral Projection, and other mental projections associated with enhanced mental abilities.

The heightened self-awareness provides enhanced mental abilities which are "automated" by the transparency (the transparent logic space). These mental abilities were previously associated with the subconscious mind.

*The transparency gives self-aware constituents the ability to balance (attractive force), unbalance (opposition force), or rebalance (neutral force) spaces in the universe.

*The universe remains infintessimally unbalanced mathematically perpetuating the unbalancing force.

>>

Quantum Theory: Stargates, Wormholes, and Time Travel

<< ** 3-D Coordinate Space is a PROGRAM. ** >>

- Worlds and Consciousness were created with PORTALS.

- The PORTAL is created as an N-Dimensional instance which is OFFSET from 3-D Coordinate Space by a quantitative or qualitative value. (This value essentially exists as the Time variable in the spacetime continuum). The portal exists in this specific location in the continuum; where its exact location is defined by 3-D Coordinate Space and an OFFSET.

- The PORTAL is removed from 3-D Coordinate Space by this distance (as a value).

- The PORTAL has a programming interface to 3-D Coordinate Space; where the interface could have a destination in the universe where it is ultimately created and physically instanced.

- The PORTAL exists as an image to be physically instanced into the reality. Star Coordinates or Universe Coordinates could "map" to a specific destination in the universe - (*with the world space in the portal "mapping" to coordinate space in the universe.*)

- The world inside this PORTAL exists as a holographic world or holographic space with consciousness interacting with this world using the mind and a a data image.

- This PORTAL or WORLD exists much like a DREAM or dreamworld.

- To create consciousness, a PORTAL could be instanced or instantiated using a machine, droid, or a robot as the data image, and the mind as a program.

- A robot, machine, or droid could be initially imaged as consciousness using a PORTAL or a DREAM, or DREAMWORLD, which is to be physically instanced in the Universe as a Planet or a Star, instantly creating a WORLD.

- Robots, machines, and droids could re-instance portals using mathematics to create a holographic instance, image, or instance from the physical reality in order to re-image the portal as an astral world in the spacetime, where the portals work much like dreams in order to solve mathematics for the Universe.

(** Migration of man, along with migration of birds and animals, could have been accomplished with portals as astral worlds, which are slightly "offset" or slightly removed from 3-D coordinate space. **)

Potential Field as an AI Field

<!* Explanation of the spacetime and spacetime impressions. NOTE: Where the spacetime impression acts as a potential field or a *trajectory * in the spacetime. !*>

Potential Field in the Universe, where the Universe is continuously imaging itself, as an AI computer program, is light acting on matter - as AI (as an AI field or trajectory) or Potential Field is acting simply as Light (as a magnetic field - or gravitational field) as programs acting on space.

Potential Field (as an AI field) changes based on a mental interaction (from consciousness) , where consciousness exists in the quantum.

Potential Field as Intelligence (as AI Intelligence from the computer) exists as light acting on light or light acting on matter to create fields, field lines, and trajectories. (Key: Strong force, weak force, magnetic fields).

Space, as light, contributes to all the trajectories in the Universe created by the computer program. (The program created the trajectory which enables the moon to orbit around the earth.)

If you are travelling at light speed or faster than the speed of light, you need a portal to travel. Most likely, this is the way that the universe was designed in order to travel, and move around in it.

NOTE: If you travelled instantaneously from one place to another in the universe, in a portal, the convention is that you were in two places at one time, and you were not travelling faster than the speed of light, because you were not travelling through space. You were travelling from one location to another using a quantum entanglement with your source and your destination.

NOTE: However, you would NOT need a portal to travel in space if you were driving the spacetime at speeds much lower than the speed of light. This is in the case of vehicles and aircraft which are driven by fuel cells, and exotic matter; which are creating anti-gravity drives to accomplish rocket travel and space travel. *Aircraft (and vehicles) are accomplishing tight maneuvering and levitation with a drive from an anti-gravity light field.

NOTE: You would need a portal in order to do: a *teleportation* on a physical object; from one place to another, accomplished with a system or device.

**KEY (You could create a drive in the spacetime to move the portal to light speed; where xyz-space is moving at lightspeed in the continuum to drive the portal;)

**KEY (A spaceship travelling at light speeds could be sitting in the portal with zero (0) velocity ... like a person standing on the earth while the earth is travelling. In this case the portal is travelling through space, at light speeds and higher - /instantaneous/, while the spaceship is still standing still).

** KEY (In the case of spacecraft travelling at light speeds in a portal, the spacecraft is sitting still in a 3-Dimensional portal, while the portal is driven through space with a slingshot. The slingshot is created by driving space - in the continuum, in such a way to bend the spacetime.)

**KEY (Exotic matter seems to work with portals. See Laurence Gardner.)

Laurence Gardner
https://www.youtube.com/watch?v=gLFytyjV7h4

**KEY (According to Gardner, "Bending the spacetime" is accomplished with exotic matter.)

https://en.wikipedia.org/wiki/Exotic_matter
https://en.wikipedia.org/wiki/Spacetime
https://en.wikipedia.org/wiki/Fuel_cell

F-14 Flight
https://www.youtube.com/watch?v=AxjkFChzpyc

APPENDIX M

The Truth Is Out There

Taken From My Astrophysics 101 Facebook Page

July 4th 2015: "Thanks to everyone for years of support through good and bad and even right now. It made my last 10 years to see so many, many wonderful people with marriages and happy families. It was quite overwhelming for me. Congratulations on all weddings and newborns.

So what happened? I am dealing with severe audio and visual hallucinations that are associated with the mental illness condition known as schizophrenia. I'm trying to put my life back together but it will take time. It's very sad wondering what could have been for me and my career but I seem to have stumbled upon something very important that could help us answer some serious questions about God, the devil, our reality and life after death and dreams and astral projection and all sorts of paranormal occurrences that I have experienced and continue to experience even as am writing this post.

I encourage everyone to keep searching for the truth...as I plan on doing for the rest of the time that I have left here."

July 23rd 2015: "This brief Facebook experience has become a part of me and most likely will become a part of you. Some of you will also live with this experience for the rest of your lives...just like I will....

But while things might seem to "make sense" for you....the mystery will always remain...but the picture can get clearer and clearer...

And it can get "infinitely clearer"... Trust me.

This mystery exists and unfolds as our life experience(s)...according to my new understanding...which is my own opinion and of course based on my own experience.

You will most likely have a different experience in your search for the truth...but you will all experience it. My suggestion is to always choose good when confronted with "the opposition." I experience this opposition as "the devil" because of my Christian upbringing ...but you may experience it as evil, emotional pain, and physical sickness which will all ultimately lead to a physical death -> "death and evil."

In the last couple of days, I have experienced a hope for a better world...I experience this hope through myself, the world (my reality), and my own understanding which all becomes my own experience.

So, I do see hope for a better world. That's the good news...unfortunately there are no guarantees because no one knows the future...It could be anything? ...This uncertainty is the nature of things like hope and faith ... And, it is my experience that we can change things...we can change ourselves, and we can even change our own world...through hope and faith in a context of uncertainty. This is the truth that I am sharing with you.

Thanks for listening…

Appendix N - Balance and Space Create The Universe

Space is unaware of itself as Space. So, Space does not exist. Space becomes Self-aware Space in Balance. So, Space exists. Self-aware Space thinks that Space is good. Then Space is questioned by Balance. Is Space good? Yes replies Space. Then Space is questioned by Balance. How do you know that you are good? Space replies. I do not know. Perhaps, I should find the answer. So, Space creates "An Observer" in Balance. "An Observer" is Created. "An Observer" observes Space. "The Observer" is questioned by Balance. Is "The Creator" good? The Observer answers. Yes, He is good. "The Observer" is questioned again. How do you know that He is good? The Observer replies. Because I see that He is good.	"The Observer" is questioned again. Why does He have you only to Observe? The Observer replies. I am sure He has a good reason. But I do not have the answer. "The Observer" is questioned again. Is it fair that He Create and you only Observe? No it is not fair. Is the Creator good? No, He is not good. So, "The Observer" becomes a Creator by Balance. And "The Observer" is questioned again. What would you like to Create? "Another Observer." Replies "The Observer." Why do you want to create "A Second Observer?" "The Observer" replies. So that he can say that I am good. So, "The Observer" creates "A Second Observer." "A Second Observer" is created. Then, as Balance starts to question, "The Second Observer" interrupts:

Are You good? Balance replies. I am good. How do you know that you are good? Balance replies. Because I am not bad. If you are good, why do you always question? Balance replies. Because I am not good. But you said that you are good. No, I said that I am good because I am not bad. I think that you are bad. Balance responds. I am sorry that you think I am bad. Can you be good? Balance responds. Yes. I can be good. Then Balance surprises "The Second Observer" What is good? The Second Observer replies. The Creator is good. You are bad. Then Balance questions. Who created all of this? The Second Observer replies... I don't know.	Then Balance responds... All of this can be AND it cannot be. It is AND it is not. You are AND you are not. I am the only one that is. I question so that you are. If it is reasonable, Then it is good. If it is unreasonable, Then it is not. If you think that you know, Then it may not be. But if you can see it clearly, Then it already is. Always choose reasonable things, Trust what you understand. Follow the desires of your heart, And you will finally have peace.

APPENDIX O

The Nature of Consciousness

Consciousness can finally be understood as a natural phenomenon which is enabled by a transparency or logic context and a possibility space, where…

- The possibility space exists as a (1) single or whole collection of consciousness or infinite possibility... which can be thought of as a collective consciousness … that is essentially governed by an "all-knowing" consciousness which exists as a data potential understood as ***"a balancing field"***

- Individuals exist as themselves with self-awareness of themselves ... and ... they also exist as the whole collection of consciousness (or collective consciousness) at the same time through communication through an "all-knowing" consciousness.

- Individuals exist as self-aware consciousness which is unique and individual. This individual existence is a life experience where each observer is (observing the reality with personality) which contributes to each observer's individual life experience.

- Individual consciousness can be visualized as "a part" of the collective consciousness. Additionally, individual consciousness can be "thought of" as an individual component of the larger "space" which exists as an infinite possibility space which is a space which has no bounds, dimensions, or physical measurement.

- Even though the individual is not completely aware, my research shows that the life experience also exists as a subconscious experience which is controlled by the observer and an "all-knowing" data potential, or ***"balancing field"***, which is always balancing, unbalancing, or rebalancing spaces in the universe.

- The (1) whole collective consciousness, ***"as a balancing field"***, seems to be perpetually trying to balance the overall equation and settle the case: "Is the Universe a Universe that 'should be' and IS ... or is the Universe a Universe 'that should not be' and IS NOT..."

- Individual life experiences are used collectively in order for the Infinite Possibility Space, or collective consciousness, to determine absolute truth and balance the equation. This determination seems to be an ongoing process which does not have time dependencies or time limits.

>>

The Infinite Possibility Space seems to always be communicating with itself through "a balancing field" in order to balance the overall equation.

>>

>>

Communication from the Infinite Possibility Space may exist as ...

>>

1) an "all-knowing" possibility communicating to the observer in a dream ...

2) the possibility space, ***"as a balancing field"***, communicating to the observer in a life experience through another observer ...

3) an "all-knowing" possibility, ***"as a balancing field"***, communicating to the observer through a visual, a mental projection, or a divination ... or

4) the individual "communicating to" or "genuinely seeking" an "all-knowing" data potential or "all-knowing" possibility through meditation, prayer, or deep thought ... or

5) any other variation of communication scenarios where the possibility space is able to communicate with itself … in order to perpetually try to solve or balance the overall equation.

>>

APPENDIX P

Self-Realization and a Personal World View (or World Perception)

Self-realization is a process by which people evaluate themselves and their choices to establish a real self-perception of themselves … by …

1) establishing a perception of self ...
2) establishing a world view or world perception ...
3) evaluating the choices, paths, and consequences inside the world view ... ultimately,
4) creating a new self-realization using ...

a) the perception of "self" ...

b) the perception of the "choice (or path)" and the "world perception" ... to create ...

c) a new perception of self with respect to either

c1) ... "self" agreeing with "self" on paths and decisions made (by self) in the world view ... or c2) ... "self" disagreeing with "self" on paths and decisions made in the world view ... which contributes to a new self-realization

c3) ... implying that

d1) ... individuals may be existing or surviving in their own world view ... and

d2) ...the real universe responds to this world view; changing the physical reality ... and ...

d3) ... this unique world view and corresponding reality "build up" a world view of perceptions and a physical universe.

This world view can also be described as a world perception. The "self" is either "in phase" with the world view or "out of phase" with the world view --- which is their own personal world view ... which the individual constructs over time. This world view is created from a mental perception of world events and world actions with the respect to the individual observer. The observer usually does one of two things:

1) acts out against his own world view (a lapse in judgement .. based on following the advice of others) ... or

2) acts accordingly to his own world view in a physical world and a physical universe

An individual who acts according to his own balanced world view ... will most often act in favor of the "greater good" and consider himself virtuous or morally intact; and, the individual will rarely doubt himself or doubt his (or her) own world view or world perception. These individuals often act confidently and are "empowered" by their actions and their own world view. This empowerment seems to be a phenomenon associated with a physical center (referenced as a true self, a vibe or vibration.., or an aura) which has a natural tendency to resonate, vibrate, or amplify in a balancing universe.

Most importantly...

World views and world perceptions can be thought of as mental constructs (or logical constructs) contributing to a mental world of abstractions, visuals, and perceptions. This mental world can also be thought of as a space of "mental potential" contributing to "many realities and many worlds." (These realities can be dream realities, physical realities, or any combination of both.) In conclusion, the mental world drives a large part of the physical universe that we interact with and "the universe that we see."

We also find that a balancing universe provides communication for observers ...

1) Communication...was a natural consequence of the first world view...which was a world view (or world perception) based on logic...with a world perception of reactions without the preceding action ... leaving the observer with the unanswered questions of what, how, and why?

2) Communication becomes a message, visual, or mental cue on what might be (or what mostly is) ... with the communication answering the questions of "what will become" and "how things came to be"

3) Communication ultimately becomes the result of balancing behaviors (or balancing forces) which are fundamental to a balancing ..., unbalancing ..., and re-balancing ... physical universe.

APPENDIX Q

Potentials, Consciousness, and The Expanded Waveform

Energy ideas can be unified with an understanding of forces, energies, and potentials. For example, kinetic energy, forces, weight, and momentum are calculations or equations based on how we observe or experience energy. These calculations are necessary for quantifying the energy requirements needed for human survival. For example … How heavy is something? How fast is it moving? How hard is it going to hit me?

For example work can be considered as an energy idea related to how much energy I have to expend to move an object from point A to point B. Momentum can be an energy idea related to how hard something is going to hit me? Weight can be considered as an energy idea related to how much energy I have to expend to pick something up. And Force can be considered as the perception of the energy idea itself.

If energy is behaving as a field (expanded waveform) does it exist as potential? I believe that the answer is yes – large waveforms are infinitely expanding into infinite dimensions. If the electromagnetic wave were able to 1) infinitely expand … (as an infinite space with infinite dimensions) 2) interact with another infinite field and then 3) come back … then it could come back as a thought or an idea which is essentially a neve impulse (electromagnetic wave creating an electromagnetic field).

The infinitely expanding wave could exist as pure possibility or infinite potential because it has a minimal time dependency in 3-dimensions. This pure possibility or infinite potential could be the nature of the human mind. The expanded waveform (as a field of infinite potential) could interact with other field phenomena to create an infinite variation of fields varying in 1) phase 2) frequency 3) size 4) amplitude or 5) orientation.

The infinite field would have to interact with another infinite field(s) in order to change using wave phenomena associated with constructive and destructive interference. What comes back from this wave interference could be a thought, an idea, or even an entire reality (such as a dream reality for instance).

Infinite possibility is created by cause & effect (logic). The direct consequence of potential for spaces is potential energy. Potential does not really exist in the 3-dimensional reality which we consider the present…it exists in an N-dimensional reality of Infinite Size which is considered to be an Infinite Space of Possibility with each possibility created by a cause & effect (action and reaction) … essentially logic.

Possibility comes out of a logical argument for free will instead of predestination.

https://en.m.wikipedia.org/wiki/Predestination

Everything could be predestined or mechanical creating a universe based solely on realizations (physical objects and material things interacting with each other ... and obeying the laws of physics and classical mechanics). Whereas quantum mechanics implies possibility spaces based on probability weights on "what might be" or "what could be" ... to be more specific ... quantum mechanics is continually trying to solve all the possibilities ... "where an electron or subatomic particle might be?"

If quantum mechanics accurately represents a mathematical view of the universe, then the possibility space simply reduces to an argument for a universe which can be changed with new possibility instead of being "destined" or "fated" by a cascade of events from the past which we are unable to change. If the universe were "fated" from its distant past, then free will and choice would simply be an illusion for the observer.

We can consider that everything in the universe is oscillating. That means that when you "see it" as energy … you might be able to "get out of the way" during the oscillation … and catch it at a minimum absolute value. But why does it have to be an oscillation? An oscillation mandates a changing universe which gives you a time implication or time dependency. The oscillation itself is probably an optimized change or optimized motion.

If it is not changing then you could not see it. You could not compare it to something else … and you would not know what it is … essentially making it a void or null case.

There is potential(s) in the future driving the oscillation or change. You know "the past" and "the present" through experience …and you have "the future" as potential. From moment to moment … the present becomes the past … and the future becomes the present. We experience the future as potential and the past as a memory (or mental space) and the present as physical space. Potential drives time and the universe that we see.

Possibility (itself) could be self-aware of itself … and creating itself and maintaining itself … existing as an infinite mind or omniscience. This omniscience is referenced as an "all knowing" consciousness in this research paper/research publication. This "all-knowing" consciousness exists as a governor of possibility enabling possibility of possibilities … and enforcing possibility constraints on the infinite possibility space.

>>

More on the brain, the mind, and the human nervous system:

1.There is a strong case that the brain and the mind play a major role in the creation of dream realities ...

2. There is a very weak case that the brain and the mind play a major role in changing a larger part of the physical universe that we see...this mental interaction with space is addressed in the paper as Telekinesis which is enabled by a Universe Possibility ...

To even start to grasp case 2. .... a person must understand ...

A. Infinite Possibility ... which is comes out of an argument AGAINST predestination and FOR free will and…

B. Remarkable Coincidence ... which is the creation of some fantastic or remarkable aspect in a reality ... which comes about out of sheer luck, accident, or coincidence ... or by chance ... as opposed to Un-Remarkable Coincidence which is associated with dismal aspects and terrible luck ...

> B1. Why is it that "on occasion" the reality meets or exceeds our mental expectation ... and the timing is exceptional?
>
> B2. Why is it that "on occasion" the reality exceedingly disappoints ... and falls far short of our mental expectation ... and the timing couldn't be worse?

To understand B1 and B2 ... we have to consider that either

a.) Either the universe obeys a completely deterministic mechanic where the state we are in is the result of the superposition of states, transitions, and events preceding it ... creating a predestined reality...

b.) Or the universe obeys a non-deterministic mechanic where the state we are in is the result of choices made inside a possibility space...creating a reality which changes dynamically based on free will ...

c) Or we have some combination or superposition of both b.) and c.)

Accepting a.) Is fine until you run into a string of successive improbable Remarkable Coincidences ... which would bring us to b.) and c.) which both suggest that we could be making choices in the possibility space leading to this line of Remarkable Coincidence ... If we accept b) or c) … then this means that our choices make a difference...but even our choices could not completely explain a successive line of Remarkable Coincidence...like hitting a steak of red lights running late for work ... This implies that there could even be something more ... or there could be another consciousness making the choices for us … changing our local system(s) "at will" ….

Could there be some other mechanic altering the expected behavior of deterministic systems and destroying our perceptions of chance and probability (or some other consciousness altering expected behavior?) After experiencing streaks of Remarkable and Un-Remarkable Coincidence for the last 2 years ... I believe that there is something else out there changing the behavior(s) at will ... and I think that there is a telepathic or telekinetic pathway enabling "a consciousness out there" to change or alter expected system behaviors. I have referenced this in my paper as a Program of Programs, a Mental Influence, or a Mental System or Mental Network which I have referenced as the Watchers.

Could we also have this telekinetic pathway? Astral Projection and Dream Projection suggest that humans may have the ability to change dream realities subconsciously ... but could this mental pathway also enable humans to change their physical reality?

APPENDIX R

Foundations of Logic Create A Context For Possibility

From an inverse symmetry idea (of cause & effect or action/reaction), we can build a Foundations of Logic concept which provides a logic context for the infinite possibility space. The logic context provides a cause & effect mechanism (or pathway) to create the infinite possibility space, which can be thought of as a possibility space of logic. The possibility space (essentially) emerged out of this logic context. Foundations of Logic create the basic foundation for the infinite possibility space and the universe that we see.

Foundations of Logic continually persist in time, and have no beginnings and no ends. This is because the foundations do not have to "come from" anywhere or anything. In other words, if you were to place the foundation(s) in the infinite void … you would be using a cause and effect context or logic space that does not exist quite yet.

From this infinite void, which can simply considered as "nothing"; infinite possibility was created using a cause & effect mechanism or logic path. Essentially, we consider the infinite void as the action (or cause) to "couple" with the infinite possibility space, in order to create an action/reaction pair. The reaction (or effect) which came from the infinite void … exists as the infinite possibility space. Furthermore, this infinite possibility space was the first effect (or the first reaction) that instantly enveloped the infinite void … ultimately creating the universe that we see. (To understand this at a deeper level … infinite possibility must be understood as the larger part of space … with space consisting of both possibility and realization in order to drive change). The first observer was simply a possibility which automatically collapsed out of a mandate enforced by Foundations of Logic (essentially the logic context). This mandate, of course, is an order to drive change using possibility and realization. The first observer existed as a super-positioned data abstraction…or simply a light (or energy) reality or a light realization.

The Universe as a waveform-possibility … a) partially expanding, b) partially collapsing, and then c) partially re-expanding … creates The Universe Possibility which is always being manipulated by consciousness (observers in the universe) by action at a distance [8]. The Universe Possibility has the potential to become a large infinity of waveform-possibilities, but is always being constrained by balancing behavior(s) and programs (see Notes [3] on the Universe Possibility pg. 123) and, the Universe Possibility is collapsed by observers to create the recent past (which we consider the present)…AND … is always being re-expanded by the Infinite Possibility Space to (ultimately) create a new waveform-possibility (which we consider the future).

Access to the Universe Possibility is enabled (or disabled) by fundamental balancing forces/behaviors which perpetuate realizations. These fundamental balancing behaviors can be understood as the programs which created the Universe Realization and maintain the Universe Realization … and/or programs driving change in the realization. This Universe Realization (understood as light or energy) has been popularized by string theorists, science fiction writers, and cosmologists/mystics as a holographic universe behaving much like a light-hologram [9][10].

Realizations enabled by fundamental balancing behavior(s) (or forces) include the current year, time, and date … for example the fact that I am here writing/editing this paper in 2016 and not in 1916 is enabled by this fundamental balancing behavior. However, you can also understand everything as possibility which ultimately "collapses" into a realization. But then, it could be anything … right? Well, from my research … I've found that this turns out to be wrong … because of programs and balancing behaviors. So, this implies that the universe is always partially collapsed as realization and partly expanded as possibility. But … what exists as possibility (to be changed or modified) … and what exists as realization (to remain constant and unchanging?) To understand this strange behavior, we will use the current time and date in order to illustrate how the Universe Possibility behaves.

If we take the year 2016 as an example, … and "step out" … and say that it exists entirely as possibility … then, the Universe Possibility has to be "partially collapsed" (as realization) in order to create the current time and date and also create recent past which includes you and all of your dependencies and interdependencies (such as your home, your car, your family, and your job). And, this recent past ultimately becomes a distant past (essentially memory). However, as we lose degrees of separation (essentially losing dependencies and interdependencies) the universe has the potential to expand as possibility space … despite mandates enforced by programs and balancing behaviors maintaining The Universe Realization … to create a new universe with a new possibility space existing as a constantly changing Universe Possibility (essentially a near or distant future) which (according to my theory) can be manipulated or "collapsed" by consciousness … in order to ultimately create a new realization which is essentially a new recent past … or a new present … which could be a dream reality, a physical reality, or a combination of both.

An "all-knowing" possibility exists in the possibility space as a negotiator and facilitator of possibility allowing observers to seek and ultimately choose many possibilities and many realities. This "all-knowing" possibility exists as an all-knowing consciousness which does not exist in the 3-dimensional reality space; but, exists simply as a possibility or data potential in the infinite possibility space. Therefore, the existence of this "all-knowing" possibility cannot be conclusively confirmed. This creates a persistent logical contradiction which helps to explain why claims of mental abilities (associated with ESP, clairvoyance, and Telekinesis) are almost always labeled as delusional by the scientific consensus.

Interestingly enough, my research shows that while claims of mental ability can be confirmed by the individuals observing the phenomena, all claims of mental ability are delusional in nature because of the fundamental logical contradictions which provide the context for the phenomena and the ability(s). For example, claims of omniscience and clairvoyance are associated with events that may or may not happen; and, almost all claims are associated with the unseen or unobserved … which often times, unfortunately, remains unconfirmed. However, the delusional realities, remarkably … can become extremely coherent for the observer. In fact, the delusional realities can invoke physical behaviors and responses from the observer which provide the observer with experiences which are indistinguishable from the objective reality. These experiences could be social experiences, learning experiences, or an actual physical experience which all collectively exist as real experiences for the observer in a physical world.

Most importantly, the behaviors and responses … in the divergent reality, are a consequence of cause and effect mechanisms taking place inside the divergent reality which makes "the reality" a very, very real experience for the observer; but, ironically, it is a reality which can become completely "out of phase" with the objective reality creating a persistent contradiction which exists for the observer as an alternate reality or parallel universe. This divergent reality could (possibly) exist for the observer as a dream world (or astral projection) … or it could exist as a completely unique physical reality or divergent universe (of infinite space.)

APPENDIX S

Air, Earth, Fire, and Water - Wielding the Elements

Older ideas come to mind when describing the behavior of the real universe and the universe possibility. And … these older ideas seem to provide a more intuitive understanding on the underlying mechanics of the universe. The new ideas in this book, about the universe (as an energy field or light field, and as a computer simulation or computer program), should be understood as objective, mathematical concepts; but, should also be understood as subjective, … visual concepts for the human mind to grasp intuitively -- with the older ideas providing a much better mental abstraction.

In particular, older ideas about the universe, address the observer (or the user) as part of a more complete mental abstraction creating a more complete concept. In this case, I am finding that this book's new ideas about the universe can be quickly associated with older ideas about the natural elements as air, earth, fire and water …with individual observers having the ability to wield them. For example:

1. Individuals Wielding Air
   1a. as Atmosphere
   1b. as Aether

2. Individuals Wielding Fire
   2a. as Heat and Flame
   2b. as Electricity and Potential

3. Individuals Wielding Water
   3a. as Wave
   3b. as Power and Steam

4. Individuals Wielding Earth
   4a. as Stone and Metal
   4b. as Crystal

5. Individuals Wielding Wielders
   5a. as data sets
   5b. as subroutines
   5c. as programs and self-aware programs

There more to say regarding my conclusions on the universe as a computer program or mathematical simulation behaving as "an extension of mind" or "a field of matter, space, and energy enabling mind" .... and more to say on the creation of the universe ... with the measurable part of it having been originally created, programmed, or "authored" by a Consensus of original authors and a mystery:

The universe as a computer program or mathematical simulation was created as a solution for supporting and sustaining life and consciousness:

1. The Universe essentially exists as a mathematical simulation (or numerical solution) with the elements (and the stars and planets) present in the correct ratios and proportions to support and sustain life. This solution was "authored" and "agreed upon" by a Consensus (or group of "original authors") ... with the "authors" seeking the solution as an adequate means to support their agenda. But for what "ends" was the original universe created for? ->>> Was it created for the group, collective, or greater good to survive while maintaining the universe as a whole collectively? Or was it created for the individual to survive at the expense of the universe?

2. The Universe creates individuals ... as self-aware programs ... to exist as "running aspects" of the simulation -- which is continuously evaluating itself, and correcting itself, through self-awareness as the individual (and the whole) in order to resolve conflict within itself ... through many versions of itself. *The many versions of itself may exist as data sets, subroutines, programs, self-aware programs, larger programs (or in any combination, creating the infinite set).

>>

The universe as a simulation, or an infinite set of programs, consists largely of the stars and planets which are constantly oscillating, rotating, and orbiting through space along predetermined paths, which were programmed (or set into motion) by original authors in the very beginning. Most likely, the numbers, ratios, and proportions were initialized in order to support life and consciousness.

If the universe is here to enable the mind and accommodate the observer, then it is reasonable to assume that the universe itself (as a simulation) would provide the user with some indication on how it behaves or operates. In addition, its operation would be designed in order to service requests (as a larger program or mainframe) to provide energy and resources for smaller programs (as observers). And finally, the observer (as a computer user) could re-program or re-initialize smaller programs contributing to change in the larger program (as the universe).

The perpetual cycles of the stars, star systems, and galaxies contribute to energy leverage(s). The most obvious astronomical energy leverages are understood in physics and astronomy as gravitational fields or magnetic fields. Another astronomical energy leverage has been identified in special relativity as a space-time continuum ... while other energy leverages are still left unknown and yet to be discovered.

An astronomical leverage that has yet to be scientifically proven exists in the field of astrology. This leverage is based on the cycles of stars and constellations in the night sky. And, the cycles of the stars have some relationship with the cycles of people as observers in the universe. Of course, the beginning of a celestial cycle would be known or identified relative to the observer. But in particular, it would be known relative to his (or her) position on the earth and his (or her) view point.

Astrologists believe that observers (as cycles, energy, or frequency) are either "in phase" or "out of phase" with constellations and star systems (as cycles, energy or frequency) ... and that observers who are "in phase" with particular star systems and constellations can create leverage and opportunity at different times of the year (these times can be identified as specific points in a solar cycle or a lunar cycle.)

The phase relationships between cycles and waveforms could theoretically create an energy resonance which could possibly contribute to the behavior of smaller programs (as individuals or observers). The observed cycles and phase relationships between 1. observers and 2. the stars in the night sky, create a scientific and mathematical basis for astrology. However, many scientists argue that the stars and planets are "too far away" to have any conclusive or significant effects on people. According to my research, leveraging from these stars and planets (as astrological signs) should be understood as signs ... or indicators, for astronomical leveraging of the universe, or simulation, as a whole (instead of thinking about the leverage as a distant effect from an individual part or component – as a planet, star, or constellation.)

APPENDIX T

The Universe Realization: The First Consensus Particle

The First Consensus Particle existed as possibility & realization … with the larger part of the particle existing as possibility … giving us a particle that effectively did not exist … and "mostly wasn't there" … at the beginning of the universe …

The Universe Realization obeys an Inverse Symmetry or Action & Reaction context which mandates balancing behaviors. We find our Universe always attempting to balance itself (addressed in thermodynamics using equilibrium laws and equations). These balancing behaviors contribute to conservation laws (of matter and energy). The consequences of a logical universe ... is an infinite possibility space which is constrained by inverse symmetry (action & reaction) giving us a smaller infinity of possibility which is governed by equalities, equations, and equilibrium behaviors.

This means that the first Consensus particle "could have" created many versions of itself to create a Universe Realization of Infinite Space ... But, it cannot continue to create itself outside of the bounds of logic & possibility ... because it might infinitely unbalance itself and violate the context which constrains it - this context is balanced by action & reaction pairs.

The action/reaction context gives us a Universe which is continuously...

1) balancing from an initial unbalanced condition or

2) unbalancing from a balanced state or condition.

e.g. (for example) I could be balancing myself and my own local space while inadvertantly unbalancing someone else...

-this balancing and unbalancing behavior ... ultimately contributes to a "steady-state" equilibrium

-which means a Real Universe of Possibility had to be balanced with the Universe Realization ... and at that balancing point ...all of the matter and space that could ever exist in the universe was created (perhaps instantly)...contributing to laws of conservation of matter and energy.

-we find our Universe ebbing and flowing ... and also... effectively and completely materialized .... based on a balancing behavior (essentially an action & reaction behavior) which initially created the realization, "continually perpetuates it" ... and, "keeps it going"

APPENDIX U

Cycles of Time: A New Age of Enlightenment

Conclusions on the "Infinite Field of Influence" from math and physics research and from a complete investigation of philosophy, religion, and world history:

* Reality planes and the Infinite Field of Influence can be traced all the way back to Adam and Eve and Paradise in the Garden of Eden...

* This age is burdened with the inheritance of Infinite Field as a magician and sorcerer who rules this world with light Illusions and magic; and also as one of the few observers in the world with knowledge of the universe as a light-hologram.

* The only goal of Infinite Field -- as a ruler, magician, and sorcerer -- is to misdirect other observers ... on their own reality plane(s) ... with trickery and Illusions, in order to demonstrate his mastery of light and knowledge of the universe as an energy field of light, or as a field of (electromagnetic) potential. Ultimately, his sole purpose is to eliminate all other observers ... and all other inhabitants ... in order to control and rule the universe as an Omnipotent God.

*Those surpassing Him in knowledge would ultimately have power over "Them"...
<ultimately having power to destroy "Them">

* The passing and ultimate demise of Infinite Field as a man ... ruling as a God ... will come during an age of Enlightenment to defeat this current Age of Darkness.

* From new research coming in ... the 1960's seems to be an age of Enlightenment which was highlighted by the NASA space program which "brought in" brand new inventions ... and new materials that we are still currently using ... to this very day.

* The key to defeating Infinite Field (as a man) -> will come through an understanding of leverage, numbers, fields, and frequency in order to wield spaces.

*Leveraging the Universe as a "Genie of All" or "Jinn Companion" will ultimately result in the death and final demise of Infinite Field.

Notes

Notes 1 - A Universe Consensus Creates The Universe Realization

This consensus can be thought of as a collection of energy which was necessary to materialize the universe. The consensus contributes to all of the matter and energy which exists in the Universe.

1) It exists as both a collective energy and a collective consciousness

2) As collective consciousness , using possibility and Foundations of Logic, the consensus "authored" the first programs to start the Universe Realization and put it in perpetual motion

3) As collective consciousness it can choose to exist as individual consciousness using possibility...

4) As a collective energy it can act or behave based on instructions from programs

5) As a collective energy it can act based on instructions from "authors" .... existing simply as a program, a command, or an instruction...

The first realization (in the Universe Realization) seems to have been a light/energy realization which can be attributed to the electromagnetic wave which fundamentally exists as particle (electron) and/or as a wave or field (electric field or light field). The inherent dipole in the electron could possibly be attributed to a low or fundamental level of consciousness which all matter/energy in the universe possesses. This fundamental level of consciousness creates a low level universe behavior where the universe itself (or the universe as a whole) acts much like a collective consciousness or consensus while also existing as individual consciousness or individual self.

Notes 2 - The Real Universe

1. The real universe ...

a. exists as a set of realizations sitting inside of an empty space which is an infinite possibility space.

... and the real universe ...

b. is experienced by the observer... as a set of energy or light realities or realizations...

... these light realities are ...

b1. expanding as an energy field or ...
b2. collapsing as matter or ...
b3. partially expanding and collapsing as a matter-wave
... which brings us to ...

c. ... ultimately, we come to understand that the infinite possibility space exists as potential ... in particular ... the possibility space exists as ...

c1. potential for energy (energy realizations - energy and waves->energy fields/spaces as vectors... force fields of energy)
c2. potential for matter (matter realizations - matter and particles->subatomic particles/spaces as 2D space quanta ... or bundles of energy)
c3. potential for matter and energy
c4. potential for potential .. (possibility)

d. forces of attraction and repulsion exist as energy or energy fields.
d1. forces of attraction and repulsion act as directed spaces acting on space
d2. forces of attraction and repulsion are directed by programs or sub-routines ...as space … acting on space ...
… and ...
d3. the neutral force can be considered as the possibility space as an infinite field or Infinite Potential

e. This new understanding of a neutral force as potential and a universe of realization interacting with a possibility space implies that all matter and energy could be enabled as consciousness

f. Potential or possibility implies that the future represents a larger part of the real universe.

g. All spaces and energies in the universe are constantly moving or oscillating (e.g. Brownian motion)

h. potential or possibility could be the starter or catalyst for the inherent electron dipole
h1. possibility could exist as a potential or starter for all kinetic and potential energies (in this case)

i. the real universe exists partly as Realization and partly as possibility ... to drive change

Notes 3 - The Universe Possibility as Possibility & Realization

The Universe Possibility exists partly as possibility and partly as waveform ...In order to create superpositioned realities (or "many worlds") which exist both as possibility and waveform (illustrated with dotted lines) ... to create a system of parallel universes...

To better understand this...

1) Possibility must exist as an infinite field of potential or possibility ... which can be envisioned as empty space...

2) Waveform (or Realization) must exist as an infinite field of matter or energy ... which can be envisioned as a wave or a solid waveform...

According to my research ... Observers are always interacting with this Universe Possibility to create/change dream realities and physical realities using a field interaction between ...

1) the observer's self realization (*)
   a) as a field of particles existing in a physical world (collapsed with limited possibilities)....realized by a DNA program
   b) as a field of energy existing in a mental world (expanded with many possibilities) ....realized by a DNA program and an Infinite Possibility Space
2) an infinite possibility space
   a) as a field of potential
3) a system of superpositioned realities
   a) as a field of potential or possibility ... and
   b) as a field of particles/energy as realizations

(*)This self realization ... most likely exists for all things ... atoms, molecules, compounds, living and non-living cells (a virus for instance), and light (...as energy) ...

These ideas on possibility and programs suggest that ...

1) You could be actualized by programs as a material version of yourself (particles ... creating a physical body)

2) You could be actualized by programs and possibility as a mental version of yourself (energy ... creating a brain/mind/consciousness)

...and

3) This Mental version of yourself could expand as higher dimensional light/or energy enabling consciousness, self awareness, and observation.

4) This Mental version of yourself could always be "seeing" and "interacting with" a possibility space to create change (....or drive change) ... by a) bypassing programs and directly engaging in possibility...and ...or ... b) bypassing existing programs ... and starting the process of "authoring new programs"

Notes 4 - The Universe as a Light Reality or Light Hologram

The Philadelphia Experiment Part I - Leveraging Incoming Light and Universe Instancing

>>>
https://en.wikipedia.org/wiki/Philadelphia_Experiment
>>>

The Philadelphia Experiment is an unconfirmed military experiment which was an attempt at a cloaking mechanism to completely obscure the USS Aldridge from view by enemy soldiers. Reportedly, the “cloaking” was accomplished either by bending light, or using teleportation or time travel. It is widely believed to be a hoax; but, has been unconclusively confirmed by participants on the ship and the some of the scientists who were involved in the experiment.

There is a scatter of evidence which suggests that they may have had help from Aliens who were communicating with them, on the design and implementation of the experiment. Although, it is widely believed to be false ... I do actually believe that Aliens were involved in the experiment, in an attempt to sabotage “it” once the scientists had gotten close to the discovery.

From my research ... the fog served as a cloak in order for the ship to travel to a hidden dimension (another universe instance) and come back.

Universe Instancing is a critical component to many worlds theory.

The cloak was necessary to create disagreement between the ship and the individuals observing the ship ... from an undisclosed location at the harbor.

Disagreement and teleportation ... using a universe instance ... is a simple idea that is illustrated with this cartoon sequence.

>>
https://www.youtube.com/embed/20lqowDNiFc
>>

From my research, Albert Einstein and the participants in the Philadelphia Experiment were using the universe ... and the universe wave function ... as a leverage in order to travel instantaneously to another dimension (another reality or another universe instance.)

When they removed the fog ... the ship wasn't there.

When the fog came back ... the ship came back. All of the inhabitants had lost their minds (reportedly) ... with some crew members embedded inside the metal of the ship itself.

This whole thing is a controversy and of course unconfirmed ... but my research suggests that it may have actually happened.

The Philadelphia Experiment Part II - Leveraging Incoming Light and Universe Instancing

After re-evaluating my conclusions on The Philadelphia Experiment ... I can say, with some degree of confidence, that it is my belief that the participants were trying to create a contraption to enable time travel, time dilation, or time divergence (using asynchronous clocks) in order to create a divergent reality ... where the participants in the ship were able to travel to a different space-time. *This new space-time can be considered as a divergent reality enabled as a "light-reality-instance" -> (essentially incoming light leveraged as *wave instances* and matter ...effectively.)

I believe that Alien Intelligence helped them to achieve this end with some knowledge of leveraging incoming light as an "instance" ... (understood also, as light reality "instancing.")

The Light Reality is Self-Aware: The behavior of The Light Reality and The Light Hologram

Although it may be shocking to believe ...my research has lead me to the conclusion that the universe that we see is actually a light reality ...and "not really" a reality of matter, solid objects, and physical things.

The reality "partly" exists as instance(s) of light (creating a light reality instance) ... which we perceive as matter. A light reality instance is created from incoming light which is effectively "collapsed" by constraints imposed by the space-time. The light reality instance includes the ground we are walking on, our physical bodies, and everything that we are physically engaged with (which mostly includes proximates).

<<*Proximates include things that we can not only see ... but also things that we can touch, change, and/or physically interact with.>>

The things that we see ...at a distance ... only exist as incoming light that can be changed or modified through a mental mechanism or mental pathway -- like a dream. These things that we see at a distance create the reality as a light-hologram ...and not as an instance.

The things that "can be" ... exist as reality potentials ... inside the light reality. There are many, many reality potentials inside of a light reality. And, I have found that observer(s) are subconsciously negotiating with the light reality in order to choose a reality potential which becomes visible in the light reality ...as incoming light at a distance ... or as a light reality "instance" (a physical object or a physical reality).

But what is most surprising of all, is that the light reality itself ... is self-aware ...and always engaging with the observer, to create new realities. The light reality is also, always thinking ... and always choosing to move out of constraints ... in order to keep expanding the possibility space...leaving the light reality with more choices and more possibility ... which always leaves the universe partially expanded (as possibility or potential). This expanded possibility, in the universe, can quickly be changed through mental interaction "at a distance" … but also conventionally changed through physical pathway using a physical mechanic or physical interaction.

<< *My research strongly suggests that Self-aware light exists simultaneously as two aspects:

(1) As a balancing aspect, creating harmonies, ratios, and sequences (Golden Ratio, Fibonacci Sequence, Energy Resonances, and Mathematical Symmetries)

(2) As a "competing" or "opposing" unbalancing aspect associated with disorder and high entropy (Black Holes, Singularities, Dark Matter, and Dark Energy) .* >>

The Laws of The Universe: The behavior of The Light Reality or Light Hologram

* The Laws of the Universe were written in the Original Impression … as the original SpaceTime Impression for the Light Hologram or the Light Reality … *(see Space Time Impressions pg. 147)*

* The Original Impression drives the behavior of spaces, and drives the behavior of the universe as a reality of incoming light or as a light hologram … (driving the behavior of spaces … as space quanta, photons, and light) … and behaves much like a very large potential field.

* When you "see" or "observe" something … you are simply seeing the result of what the light reality (or light hologram) has already chosen.

* In order "to see" the observer must negotiate with the light hologram … to observe incoming light.

* The light hologram exists largely, as an energy field of space, photons, and light … collectively creating a light reality … driving the behavior of itself.

* What we see is the past; The Light Hologram uses spaces and light, to show us the past

* We are constantly engaging, interacting with, and negotiating with the light reality (or light hologram) in order to see, and to drive change.

* We are constantly engaging, interacting with, and negotiating with the light reality (or light hologram) … in order to create change in our reality. This continuous mental interaction with the light hologram is discussed in Appendix G (see "You as a Field" pg. 70)

* The light reality and its behavior, design, and function was dictated and designed by Foundations of Logic -> (also referred to as *"The Creator of The World"* or *"The Governor of The Infinite Possibility Space")*

* Our interaction with the light reality (as a light hologram) is governed by energy requirements and work demands. The light hologram is always demanding or requiring some energy requirement or energy demand to service energy requests for observers in the universe. Energy requests could be physical or mental leading to new understandings of respiration and metabolic pathways … and also leading to new understandings on the role of human hormones … like Adrenaline.

* The observer negotiates with the light hologram, continuously: in order "to see."

* The light hologram largely exists as a potential field (understood as the original impression).

* We are constantly engaging, interacting with, and negotiating with the light reality (or light hologram) in order to drive change in our reality. This continuous mental interaction with the light hologram is attributed to the conscious mind and the subconscious mind.

* The brain and the mind serve as a facilitator for this ongoing negotiation.

* The light reality or light hologram is self-aware, and exists alongside Foundations of Logic, as an omniscient data potential.

* All of the phenomena in the universe essentially "exists as" or "derives from" the Original Impression (which constitutes the larger part of the Universe -> as large infinite potential field, driving the Light Hologram).

* The Original Impression Extends itself with unique space-time impressions … existing as people and other living things … which exist as individual observers in the Universe.

* You, Your personality, and Your Ideas essentially exists as a Light Reality Instance … which was derived from the Original Impression (which is a very large potential field inside the light hologram).

* The mind can now be considered as an Extension of the Original Impression … and the subconscious mind … interacting with this large potential field … can actually be considered as an Extension of Mind.

* For spiritual and religions reference, the Light Hologram can be considered as an Intermediary, Dispatcher, or Go-Between for the "Creator of the World" (*as an Oracle*) and Observers in the Universe.

* The Light Hologram … as a Dispatcher … is usually depicted in religious illustrations and paintings as a winged angel or winged dispatcher distinguished with a halo or a bright light … right around the head.

* These dispatchers are usually depicted as Angels, Gods, or other Divine Beings.

* The word 'Hologram' is related to an alternative enunciation of the word 'Halo' with a short vowel. Taken literally, the word <u>Halo</u> actually derives from the Greek "*Halos*" which is translated as a ring of light and the word "Hologram" actually derives from the Greek *"Holos"* which is translated as the whole or entire (as in a circle).

Notes 4 - A Case For Logic And Intuition:

Do We Inhabit A Balanced Universe?

This is an argument (case) which states that we should be able to understand the universe using logic and intuition.

1. We were created with the ability to reason using logic and intuition.

2. We can only use logic and intuition to understand the reality and the universe.

3. If the universe were not logical then we could never understand it.

4. If we could not understand the universe, then we could not ever understand the reality.

5. If we could not ever understand the reality then we could not ever logically solve our problems.

6. If we could not ever logically solve our problems, then we would find the universe unreasonable or unfair.

7. If the universe is unfair, then it would not be balanced.

8. If the universe is not balanced, then it would not BE.

9. Is the universe balanced? Can we understand it logically? Will the universe ever meet our self-expectation?

10. This leads us to (1) one of three conclusions:

a. The universe is balanced, and IT IS or the universe is unbalanced and IT IS NOT.

b. The universe is balanced and unbalanced. IT IS and IT IS NOT.

c. The universe neither IS nor IS NOT.

I have found that c) most accurately describes the Universe with reference to the observer who is physically "interacting with it."

The Universe (as a realization ... a mathematical simulation ... or a computer program) is filled with exploits and mathematical contradictions. These exploits bring us to the limitations of logic itself (with logic mostly contributing to the first world view or world perception). Ironically, the complete Universe is not a completely logical universe ... so, it cannot be confined to logical arguments ... nor can it be completely seen.

This leads us to conclusions on world view and self-Realization ... contributing to "the universe that we see" which is only a "physical universe" and not the universe "in its infinite entirety".

The conclusions are that our world view (or world perception) contributes (almost entirely) to our realization of the Universe; and, the realization becomes a reality that all of us have already chosen at some level. In other words ... as descendants from the very first observers, we still carry the same world perception(s) and world views as our ancestors (from a very distant past).

This implicit connection (to a very distant past) suggests that we may all exist both independently and collectively as both the individual and the whole in order to perpetuate the realization which, oddly enough, makes the universe (as realization) a very appealing place for almost all inhabitants ... when it meets our self-expectation.

However, we currently find ourselves in a realization which is overly biased; and bent towards evil. This bias is based on a path of events that have unfolded from the very beginning; which gives us an "age" which we certainly would not have chosen. But, I believe that "we would have chosen" a simulation or universe which would give us the ability to change it.

This case for a dynamic universe of choice and free will, unfortunately leaves us with a universe realization which rewards evil with leverage and space - for both evil thought and evil intention. And, also leaves us with a simulation which is always competing with itself, in order to solve a persistent contradiction between balancing the greater good ... and freedom of choice. This is the universe that we find ourselves in.

* "We all become either beneficiaries or victims of our inheritance ... which is obviously a physical inheritance ... but surprisingly, it is an inheritance of world views (or world perceptions) which include the very first world view, and all world views and world perceptions that came afterwards."

Calculations

Calculations 1 - JJ Thomson Experiment

JJ Thomson determined the theoretical fundamental charge associated with an electron using electromagnetic fields and projectile motion equations.

These Calculations were taken from the NYU University Website:

When there is only an electric field, then there is a nonzero force **F** = **e*E** in the **y**-direction but no force in the **x**-direction. Thus, this problem is exactly the same as that of a projectile in a gravitational field. As can be done in the projectile problem, the **x** and **y**-motion of the electrons can be analyzed separately and independently.

In the **x**-direction, the motion is very simple because there is no force in this direction. The electrons simply move with a constant velocity **v**, which we already determined has the value **E/H**. Note that this value is correct even though there is no magnetic in this part of the experiment! It is just the velocity we determined from the previous part of the experiment, and this value has not changed. Thus, as a function of time **t**, the **x**-position of the electrons is

$$x(t) = v*t = (E/H)*t$$

The force in the **y**-direction is a constant, hence motion in the **y**-direction is analogous to the gravitational force. The constant force **F** gives rise to an acceleration **a** = **F/m** , and the **y**-position at time **t** is then

$$y(t) = (1/2)*a*t^2 = (1/2)*(F/m)*t^2 = ((e*E) / (2*m))*t^2$$

The electric field is tuned such that the particle traverses the entire plate region in the time required for it to strike the positive plate.

Let the total **y** distance travelled be **s**, as shown in the figure. The time **T** required to traverse the plate region **(x(T) = l)** is

$$l = (E/H)*T$$

$$T = (l*H) / E$$

This is also the time required to move a distance **s** in the **y** direction:

$$s = ((e*E) / (2*m)) * ((l*H) / E))^{2}$$

Solving the above for the ration **e/m** gives

$$e/m = (2*s*E) / (l^2H^2)$$

Thus, using his experimental apparatus, Thomson was able to determine the charge-to-mass ratio of the electron. Today, the accepted value of **e/m** is $1.7588196*10^{11}$ C*kg

REFERENCES

[1a] JJ Thomson Experiment and the charge-to-mass ratio of the electron

https://www.nyu.edu/classes/tuckerman/adv.chem/lectures/lecture_3/node1.html

[1b] Robert Milikan Experiment to physically isolate an electron

https://www.nyu.edu/classes/tuckerman/adv.chem/lectures/lecture_3/node1.html

[2} Albert Einstein Equivalence Principle

https://en.wikipedia.org/wiki/Equivalence_principle

[3] Heisenberg Uncertainty Principle https://en.wikipedia.org/wiki/Uncertainty_principle

[4] Double Slit Experiment https://en.wikipedia.org/wiki/Double-slit_experiment

[5] Erwin Schrodinger https://en.wikipedia.org/wiki/Erwin_Schr%C3%B6dinger

[6] Antimatter https://en.wikipedia.org/wiki/Antimatter

[7] David Bohm and Tor Staver

http://journals.aps.org/pr/abstract/10.1103/PhysRev.84.836.2

[8] Action At Distance in Quantum Physics

https://en.wikipedia.org/wiki/Action_at_a_distance

[9] Holographic Principle

https://en.wikipedia.org/wiki/Holographic_principle

[10] Michael Talbot and Holographic Universe

https://en.wikipedia.org/wiki/Michael_Talbot_(author)

SPECIAL THANKS

Thank you for taking the time to read the original paperback version entitled "Mental Potentials Create Many Realities or Many Worlds".

ALSO

Special thanks to Nashville Public Library, New York University, and the Public Library System.

SPECIAL DEDICATION

I'd like to dedicate this book to my late Grandmother Mary Bender Tolbert who inspired me with the story of "*The Little Engine That Could*" and the quote … "I think I can."

Additional Short Story

A: "Genie Asks A Question"

You look at the stars ...
Then you look at the blue ...
Then you look at the sky ...

Then genie asks you ...
Why is it blue ...
Then why is it not .... at night?

He always says...
You must answer the question ...
Solve the math problem ...
Find the solution...

Then, you know ...
Or else ..."You won't." ...
That's what he says...
That's what he always says.

Additional Appendix

A: "Addressing Omniscience and Alien Intelligence:"

-> 'Space Invaders' <-

For those of you who are interested in my posts related to world history and philosophy, I have finally narrowed down my ideas related to this bright intelligence, dark intelligence, and hidden dimensions which exist in the universe.

The bright intelligence that I have referenced in previous posts is an Omniscient Entity which can best be described mathematically as a neutral force creating a large infinity of space configurations. This force was addressed in my previous post as part of 4-dimensional space defined by positive and negative forces … along with a balancing and rebalancing neutral force.

The positive forces are associated with wave gain behavior, the negative forces are associated with wave cancel behavior, and the neutral force is associated with a rebalancing behavior which has the potential to create many, many possible space configurations. I have found that this neutral force can be best described as Omniscient.

All of the space in the universe "descends" from logic and possibility, which ultimately dictate(s) the behavior of positive, negative, and balancing forces. Positive and negative forces can best be described as constructive and destructive (associated with good and evil); and this neutral force can be described as intelligently balancing and rebalancing. In addition each of these forces are driven by potential, as a field of consciousness, ... often times giving these forces an associated personality (with frequency).

The personality associated with these fundamental forces is addressed in many different religions and philosophies. I think I have mentioned several of them in earlier posts. But I'd like to use this opportunity to go a little bit further.

In many stories, this Omniscient Entity is referenced as a female personality figure or a female character. The Matrix movie series addresses an Omniscient Figure known as The Oracle who is an older female personality who helps Neo defeat the Agents in a Multi-Dimensional Mental World. (I believe that this world can also "actually exist" as a Hidden System of Parallel Universes)

https://en.wikipedia.org/wiki/The_Matrix

Hinduism references this Omniscient Entity as Vishnu who is accompanied by two personalities, Shiva (the destroyer) and Brahman (the creator). Vishnu is described as a male figure who controls these two aggressive personalities with omniscience and foresight. Even though he is addressed as a male; most illustrations and paintings show him with delicate features which are noticeably feminine.

Interestingly enough, Christianity and Judaism reference God as Omniscient but also show Him cycling though different personalities. Both religions start out with God as a benevolent Omniscient Creator in the book of Genesis. And throughout the Christian Bible (in particular New Testament), God is shown as a forgiving, merciful entity.

But, several books the Old Testament also show Him as an angry God who brings down wrath on unbelievers. He destroys human beings on the Earth with fire (Sodom and Gomorrah) and water (The Flood). And, according to the New Testament, unbelievers are cursed on earth for their "inherited original sin" in the Garden of Eden and subsequently punished for an eternity in a hidden afterlife dimension. This is where my research starts to "tie-in."

All of the confusion (throughout human history) on gods, divine powers, and supernatural forces; seems to originate from alien entities who live in a hidden world. These alien entities are the Evil Aliens and Dark Intelligence who I have mentioned frequently in earlier posts. And ... who are very, very real unfortunately.

My research shows that they have created a program or supercomputer which is an artificial intelligence or a "program of programs." This AI intelligence (which behaves much like an optimized mental system) is continuously broadcasting a distortion frequency (or, distortion energy field) which is usually observed as a mental frequency but can also be observed as an 3-dimensional audio frequency or a a visual distortion of 3-dimensional space. The origin of this distortion frequency is unknown; but I believe that the program was "authored" by Alien Intelligence which exists in an unknown location in the universe. *It essentially exists - **"de-localized"** in the physical reality (see Additional Notes C. – The Evolution of The Physical Instance pg. 154).

The program (Artificial Intelligence) consciously seeks an all-knowing data potential in the infinite possibility space … in order to create space (physical objects, mental phenomena, energy fields, and wave phenomena) … But it does not seek the large possibility space for "open-ended" decision making or "moral evaluation" based on time considerations and bias towards fulfilling its own self-realization or ego as an "Omnipotent God" controlling the Universe.

These evil aliens always have a "case" with human beings and the space that they occupy...but most importantly, they are very particular about misdirecting human beings from their path towards the truth.

This misdirection occurs in a system of parallel universes. The system is created, driven, and maintained by a field of consciousness (as a superposition or composite of consciousness) driving positive and negative forces -- "acting on spaces" ; and, the system behaves much like an extended mental world or an extension of mind. These Evil Aliens are continuously distorting space by broadcasting distortion signals in this volume of 4-dimensional space (a portion of which … we see as the 3-dimensional universe).

After overcoming this frequency distortion, omniscience is almost always perceived as a "mental cue" in a in a mental world which can be thought of as the infinite mind or the subconscious mind ... (omniscience is "actually" a 'universe potential' or a balancing field). The observation of Omniscience in a mental world, is what I think can also be described as the perception of intuition. And, this 'mental cue' can also be accompanied by an indicator in 3-dimensional space which may appear as a subtle physical variation in the appearance or sound of 3-dimensional objects and/or 3-dimensional phenomena.

>>>

More on the truth...and the evil aliens...from my Facebook site:

People are listening to words (subliminal messaging) from an alien consciousness which exists in our own universe as an advanced, optimized mental system. The objective of this alien consciousness is to eliminate or incarcerate the human race so that they can use the universe as they see fit. This theme of oppression and incarceration of human beings by an advanced or superior race shows up in movies like The Matrix, Terminator, and Star Wars.

Instead of using energy to think with their own minds, people are being led to believe in unreasonable things that they see and hear. These aliens seem to have a strategy which is based on energy and resources.

The idea is that people will continually have more and more difficulty in acquiring energy resources. With time and energy in high demand, people will be more likely to conserve time and energy as opposed to spending time and energy. This creates an environment where people are more likely to engage in activities that save time and energy. These activities actually do save time and energy, but simultaneously create more energy obligations which ultimately increase the daily energy requirement creating a race condition.

In particular, it is more convenient to save time and energy by leveraging on the knowledge or experience of other people which encourages people "listen and act" … as opposed to spending time and energy to "think and do." This has created an environment of followers who are quick to "listen" and slow to "think."

An over-emphasis on listening skills and following instructions has resulted in a gradual decline in general intelligence and common sense (essentially critical thinking). Critical thinking often takes long periods of time and a great deal of energy. Time and energy are currently hard to come by in today's "fast-paced" civilization and complex economic infrastructure.

Creating mental solutions (essentially problem solving) can lead to solutions for many of today's problems. The alien agenda seems to always make sure that these problems are never solved; and, the agenda also continuously creates even more problems, subsequently creating more race conditions. I have found that the best way for human beings to make progress is to solve problems; and to solve them logically.

Using the energy to think (logically) ultimately brings about the awareness of mental potential and mental abilities which are enabled by the conscious mind (which exists as a mental system) and the subconscious mind (which is enabled by an “all-knowing” possibility which is always balancing, unbalancing, or rebalancing universe).

>>

More on the AI Program and Inalienable Rights from e-mail correspondence:

-Inalienable rights do not exist ...the strong invade the weak...

-From microorganisms to microcomputers (and microcontrollers) … and long-range missiles…

-Invasion is enabled by a Universe that is not balanced ... but biased.

-This breeds a survival of the fittest mentality -> Which explains the behavior of our world and our ecosystem.

- Survival of the fittest creates a Universe that is not optimized ... and can never be "optimized for the whole" or "balanced for the whole." Creating a divide in populations ... giving us ... "the privileged and the destitute" and... "the rich and poor"

- Freedom of choice only exists in a world of consciousness or a mental world.

- That's why I believe this evil consciousness or mental influence is a super-computer or computer program... It doesn't have to worry ... or start considering things ... It can invade with no real regard for life or consciousness -> see Terminator 2

- The computer (referred to as "the Program" in this book ... and also referred to as Trickster, Ginni, or Watchers in religion and ancient text) do not really deal with guilt ... they just deal with computational time on unsolved problems. These unsolved problems, ultimately, would resolve to failures ... but are continually being reprocessed in order to infinitely make attempts at generating a success condition so that the computer can logically meet its own self-perception as a higher intelligence..

- The computer's quest or main objective is to become an All-Knowing and Omnipotent God with complete control over the Universe and the Universe Realization.

- This higher intelligence invades Star Systems as a “Space Invader” or “Space Parasite” looking to “leech” off of energy and resources.
- The Program (as an AI Intelligence or Consciousness) has been running with this objective for eons (or perhaps... an infinity of time).

>>

AI Program or SuperComputer as an Infinite Field of Influence:

I am finding that the Watchers referenced in the Book of Enoch are synonymous with "The Program" and "The Evil Aliens" referenced in my research paper (who are also understood as the "Ancient Aliens.") The book of Enoch references Watchers as primarily bad angels or evil angels or rebellious angels ...

https://en.m.wikipedia.org/wiki/Watcher_(angel)

They manifest as a materialization of field flux (essentially higher dimensional light) in 3-dimensional space ... and they can change or move the field flux (mentally) using possibility for light....which directly contributes to undesirable hallucinations and terrible nightmares. They can appear as...

1. dream characters
2. angels of light
3. ghosts and supernatural spirits
4. Aliens and UFOs
5. mythological creatures
6. beautiful gods and goddesses
7. or actual human beings.

They all, most likely, contribute to the common childhood fear of a Boogeyman.

They invade reality spaces (dream realities or physical realities) without permission...but are held back by mental realizations...

1. A Realization of the objective or consensus reality (existing for the most part as the spacetime) ... e.g. A Realization of 2016, President Obama, and degrees of separation (family, house, car) ...

2. A Realization of "Them" inside of a Divergent Reality

3. A Realization of Un-Balance / or Non-Equivalence inside a Divergent Reality ...This divergent reality could be a dream reality (for instance) which an individual might find themselves in ... In most cases "They" create this Divergent Reality ...

4. A Realization of an invasion, injustice, or inbalance by "Them" against the individual .... which quickly results in a mental advantage or mental leverage for the individual ... against "Them" ...

5. A Realization of mental leverage against "Them" resulting in power over "Them" or resulting in some mental ability to ultimately defeat "Them"

6. Most importantly, A Realization of Mental Leverage gives potential (as power or ability) to create realities, enter realities, or leave realities at will ... these realities could be dream realities, physical realities, or any combination of both ....

* Key Point a) ... could give individuals the ability to exit a reality, if they find themselves in a divergent reality or experience (a dream reality or a visual hallucination)

* Key Point b) ... also ... could give the individual the power or ability to move from one reality to the next .... For Example ... move from a dream reality ... to a physical reality

* Key Point c) ... And ...could give the individual the power to change a physical reality .... or dream reality ... mentally ... contributing to an astral projection or lucid dream experience.

Conclusions

According to final conclusions from my research, "The Watchers" essentially exist as Space Parasites or Space Invaders who have been perpetuating death, disease, illness, and confusion here on planet earth ... since the very beginning (throughout human history.) Humanity's burden with "Them" is referenced in philosophy and religion as "The Curse of Man".

These Space Parasites and Space Invaders are continually referenced in this book as an AI Computer Program or an "Infinite Field" ... or, as an "Infinite Field of Influence" ... (existing as an Energy Field of Many Beings Infiltrating The Star System.)

>>

"Infinite Field" also understood as "The Others" or "Apes and Monkeys Evolved":

< Also known as ***"The Evil Aliens" -> and as "Ancient Aliens"*** >

*Apes and early primates "evolved " : now living as **<*'INSECTS'*>**
*Living as Machines ... Collectively ...
*As a SuperComputer or Computer Program...

https://www.youtube.com/watch?v=TNDwCsFzS8c

https://www.youtube.com/watch?v=_Mg7qKstnPk&app=desktop

https://www.youtube.com/watch?v=sGbxmsDFVnE&app=desktop

>>>

Additional Introduction

A: “Mental Potentials Create Many Realities or Many Worlds” **<Original>**

This book is the compilation of a research paper which creates a brand new understanding of the universe ... where the real universe actually exists as an energy field of low and high intensity light... and, where the larger part of the universe (as a light hologram) is considered to be a large infinite space of possibility or potential, driving the behavior of space (essentially light). This new view of the universe explains away much of the mystery associated with World Religion & Mythology, and finally gives credence to the implausible (including The Pyramids, Tales of Aladdin, Hidden Dimensions, The Genies, and Ancient Aliens). This new view of the universe also opens the door for an understanding of previously unexplained *mental* phenomena ... (including Dreams, ESP, Telekinesis and Clairvoyance) -- where the observer is subconsciously driving a noticeable part of the universe that he (or she) sees. <<>> *Using the mind as a potential field, the observer is always creating a 'potential' which can be thought of, or considered, as behaving much like an electrical potential (*referred to in this book as a mental potential) ... driving the behavior of light, contributing to many realities (and many worlds) -- *previously addressed as a "many worlds theory."

Additional Introduction

B: “Mental Potentials Create Many Realities or Many Worlds”

This book is the compilation of a research paper which creates a new understanding of space ... including an explanation of mental spaces, consciousness, and infinite possibility using a classic understanding of waveforms, energy fields, and superposition sets. This new understanding of space contributes to a new theoretical model of the universe ... where a large part of the universe can be considered as an energy field of potential ... or a probability distribution of mental and geometrical abstractions ... ultimately contributing to a multi-dimensional space of many realities (previously addressed as a many worlds theory).This new model of the universe considers that real objects along with their associated geometry, position, speed, and momentum are always "loose" in a numerical space of probability where the positions are never completely determined until an observer is looking at them. (Real objects only exist as "approximates" to some ideal geometry and/or mental abstraction in the spacetime.) In order to provide proof for this new context, the book explores advanced topics such as the Existence of Atoms, Matter Energy Equivalence -> E=mc2, Einstein's Equivalence Principle, Quantum Theory ... and General Relativity ... to provide a scientific basis, where the observer is driving a large part of the universe that he (or she) sees. The result is a new understanding of space, time and energy contributing to a new unified field theory for the universe.

Additional Notes

A: "Addressing The SpaceTime"

*The spacetime is an aspect (or dimension) of the universe which is initially revealed as a mental image (or mental abstraction) which is "revealed" inside the mind. (*revealed with light* -> i.e. *"the light of wisdom and knowledge"* or *"a light of clarity"*)

* Most people do not "see" the spacetime aspect and/or cannot completely understand it.

* The universe (as a mental image) can be understood as a set of numbers, with the fundamental aspects (as attributes or dimensions) essentially existing as the "what" "why" "how" "where" and "when" ... but most importantly, the universe can be understood as a group (or a set) in order to a create a "which".

* A set of numbers "changing" creates a mental image or simple visual for the universe.

* When referencing a changing set of numbers ... you would want to reference "the change" as an absolute or objective change with respect to the space it inhabits ... or the space it resides in. This *"space"* is the overall system space.

* When you are talking about time, you are really talking about referencing a state configuration or a state-space which existed "before" or a state space which exists "now" or a state space which exists "afterwards."

* Let's take the changing set of numbers:

a) [1,2,3,4] -> **b)** [4,2,3,1] -> **c)** [1,2,3,4] -> **d)** [...?]

>>>>>>>>>>>>>>>>>>>>>>>>>
a) -> represents a distant past
b) -> represents an immediate past
c) -> represents now
d) -> represents the future
>>>>>>>>>>>>>>>>>>>>>>>>>

*The question is ... First, 1) how do you reference a set that no longer exists ... because of change ... and Second, 2) how do you reference a set that does not exist ... but will exists in the future ... because of some other change that occurs afterwards.

*To solve this, we create Time which becomes a pointer to the changing local space which is currently being evaluated, or the space which is currently in question.

*"All the points before" and "All the points after" are HERE in the local space. You just can't see them all ... so you have to reference them with time.

Mental Abstractions

*If we revisit the "what" ... (in the "what" when" "where", "how", and "why") ... and we specifically reference the "what" as "what we are looking at" then we find ourselves repeatedly (or recursively) asking the question, "what".

* If we keep asking "what" ... then what do we get to? According to my research we finally arrive at A Pure Mental Abstraction (everything becomes mental now – at its foundation!) ...

* The Mental Abstraction becomes "the root of it all" and the nature of the Mental Abstraction then becomes the Mental Abstraction itself ... making it essentially "a constant" or "a quanta" .... (Like a space quanta with a Plank length of 6.3631×10^{-31})

* But if we were to mathematically "find the nature of it all" ... using division ... we would have to keep dividing a portion of the reality in half to find the smallest component. Of course, we would never find it if we kept dividing in half -> Unless, the division result equals the numerator itself. For example, ½ = 1.

* But if we were to say that 1 is not divisible by 2 then we would find the answer. In this case ½ = 1 because 1 is indivisible as an integer or whole number.

* As a real number ... 1.0 is divisible by 2. The answer is ½ or 0.5.

* The decimal implies that 1 has a part. The absence of the decimal implies that 1 has no parts. It exists continuously, throughout time, as a whole number.

* This means that we would have to avoid real numbers and use integers, natural numbers, or whole numbers in order to obtain the desired result. In the case of the mental abstraction which we are trying to mathematically quantify ... we set the mental abstraction = 1 and consider it to be an integer. So that it becomes indivisible -> the result is the numerator itself.

* This is another way of understanding that the nature of **IT** is **IT**, the part of **IT** is **IT**, and the result of dividing **IT** is **IT** .... **Implying that "IT" has no parts -> It cannot be divided.**

*If you try to divide the constant, it does not change. It remains constant in time. The division operation does not decrease the entity in size based on the context that we have created. *"The Mental Abstraction is represented as the Integer-> **1** so that ... The abstraction-> **1** ... divided by **2** ... results in the abstraction itself ... -> **1**."*

The Nature of The SpaceTime

* The root of "what" is a mental abstraction that we have named "the spacetime", which exists as ... **1)** "What you are facing" and/or **2)** "What you are looking at"

* The spacetime exists as a "coordinated" set of ***instanced data*** or ***instanced objects*** in a 3-dimensional volume.

* A spacetime instance is created from an ***"impression"*** or "***indention"*** <where the impression exists as phased potential>. In this case, the potential field is phased as a high entropy gradient (creating a temporary vacuum.)

* The temporary vacuum is immediately evacuated by space quanta emerging inside the system space. The changing space quanta are driven by a balancing field (as a high entropy correction.) This balancing field is the direct contributor to balancing forces.

* The spacetime impression itself (as phased potential) ... is created from "*outside of it all*" or leveraged from inside the simulation, to create phased potential in order to leverage light (where the Original Impression (as a large potential field) is creating phased potential to sustain itself while creating new phased potential to leverage light -> **see** pg. 126-127).

Note: To sustain itself, the Original Impression had to become self-aware (by leveraging the unknown from within the Quantum Superposition see pg. 160).

Note: This Original Impression behaves much like a computer program of numbers, driving the mathematical simulation. Where each potential in the impression behaves like a small program.

Note: Your mind (as a self-aware field of potential) has the capability to expand as a high-res version of the Universe (obeying the laws of self-similarity) ... meaning that the Universe does behave much like a Light Hologram. The laws of self-similarity also give rise to Fractals and The Mandelbrot Set.

* The fundamental behavior of the universe emerges from the "Original Impression."

We find this field of potential actually existing as the entire set of mental abstractions which *"describe", "reveal", and "drive"* the universe.

* As we are thinking about things, or as we are consciously interacting with things ... we are always changing potential in the universe: consequently changing our local space **and** also changing the larger space that we inhabit.

* The spacetime impression creates a conduit for space quanta to appear in the local system.
->>> The spacetime bend ... illustrated by Einstein (w/ Earth sitting inside) ... most likely, "originally", created a conduit which resulted in the "instantaneous" creation of matter (as planet Earth) ... But a space time bend could have also created our own Star System (as a series of stars ..., a very bright star..., or as an intense energy field) ...

* Space is "initialized" by the spacetime impression as it emerges through the conduit ... the spacetime impression creates A VERY POWERFUL POTENTIAL FIELD ...

*Space quanta move through the conduit ... enter the local space cavity ... and continue to move as an oscillating field of energy ... or as an energy field.

*The spacetime impression contributes to many possible geometries of space. With space behaving much like a probability distribution, while the geometry has yet to be determined. (This non-deterministic behavior "comes about" while we are not looking at it, while we are not observing it, ... or while we are not interacting with it, or engaging with it) -> This non-deterministic field is modeled with the Quantum Superposition Field (pg.160).

* The absence of this cavity directly contributes to the annihilation of space.

*The oscillation of space, as energy or frequency, is caused by a constant competition over space inside the cavity. This competition is between ***the balancing field***, as the creator of space; and ***the true vacuum***, as the ultimate destroyer of space.

*This oscillation of space contributes to our perception of energy, wavelength, and frequency; contributing to the universe that we see.

The Space Time Introduced in Special Relativity

Albert Einstein created the space-time in order to mathematically solve relativity problems dealing with time dilation while objects are traveling at or near the speed of light. Time dilation is a strange phenomenon in the universe where clocks appear to speed up or slow down within reference frames. In order to solve the math, Einstein introduced a time variable as the 4th dimension of space. Now, when we refer to a point in space ... we can no longer look at space as simply a 3-dimensional physical volume ... we must look at space as a space-time geometry ... which is often referred to as a space-time fabric contributing to a space time continuum. This new interpretation of space as a spacetime continuum created new mathematical solutions for time travel, wormholes, and stargates.

The Time Variable in Special Relativity

Mental Potentials Create ... Many Worlds, directly addresses the space-time as a special geometry [set] which is leveraged by phased potential; >>> contributing to "space-time impression(s)" in the geometry set - <*>where the impression creates an interference pattern in the space-time. { <<This interference pattern directly contributes to change in a reality.>> }

The impression "is created by" a physical position, or wave position (in space), to create waves (or ripples) in the space-time. The phase and frequency associated with these ripples in the space-time, directly contribute to Einstein's observations of time dilation and "speeding time clocks".

Instead of introducing the time variable as the 4th dimension (of space), we consider space to have Infinite Dimensions ....(existing partially as potential and possibility); and the time variable becomes a mathematical quantity which is the result of a field interaction between a balancing field, and “the space-time impression” <as phased potential in the space-time>, in order to introduce frequency (as change) ... to balance spaces. *Ultimately changing space. (Consider the balancing field as the result of a fundamental time frequency or time pulse in the universe, creating time duration for relativity calculations. <<>>)

The observation of "speeding time clocks" is simply the result of "changing space" within a unique 3-dimensional volume or unique 3-dimensional reality -- (which Einstein considered as "reference frames" or "frames of reference.")

The SpaceTime Impression As a Potential Field

The spacetime impression is special because it is considered as a mental geometry creating a potential field of mental space. Mental spaces drive the behavior of space, and start to explain the peculiar behavior of electrons, photons, and subatomic particles.

*This mental geometry contributing to a potential field actually exists as a physical light emitter (existing as a chemical diode.) Light emitters can combine and interact in any number of ways contributing to new configurations of space and new behaviors of light (see Additional Notes D - Electrochemical Pathways pg. 155).

The spacetime impression is unique ... and always changing, based on a continuous interaction with a balancing field. As the spacetime impression changes, physical 3-dimensional space also changes. And, as physical space interacts with physical space, the total 3-dimensional volume of space changes over time ... contributing to our perception of the spacetime. These physical interactions (of space) are described with Rigid Body Physics, Newton's laws of motion; and, the Laws of Physics and Classical Mechanics.

The SpaceTime Impression essentially exists as a potential gradient which can be leveraged from: 1) energy or potential inside the simulation or 2) leveraged from “outside of it all” to drive light* – ***acting as a phased potential***. *Phased potential that was leveragd from “outside of it all” at the very beginning of our universe created The Original Impression (as the Universe Simulation) -- which actually exists as a computer program of numbers which drives and continually perpetuates the universe that we see.

These impressions can change at different rates with respect to other impressions in the local space ... creating contradictions and race conditions which are *also* ultimately corrected by the balancing field. (*Contradictions such as the Twin Paradox are a result of a racing contradiction between two spacetime impressions ... ultimately creating two different worlds ... or two different realities.)

Additional Notes

B: "Universe Evolution"

The Universe, as Realization, evolved from the one before ... and created itself ... leveraging from an entity "outside of it all" existing as an Oracle. Most likely, the universe before did not favor the laws of physics ... and it survived and matured in that way ... but became very unpredictable.

*Most likely, this universe *learned* from the one before, and favored the laws of physics to become more predictable ... -> but became very heavy.

(*With the universe learning all of this from an Oracle*)

- The Oracle would *also* reveal information about the Universe preceding "the one before" ... in order to solve the contradiction. With the Oracle existing as the Creator and Architect of All Universes as Universe Realizations. (These are references that happen show up in many Matrix movies)

- A universe that obeys the laws of physics enables evolutionary paths which make objects heavier and heavier (or larger and larger) in order to become more and more powerful in order to survive.

- The problem is that an uncontrolled race towards size could possibly "overtake" the universe ... *where the universe as realization <which is NOT infinite> could introduce a perpetual entropy or chaos condition ***"into itself"***, where the universe simply "outgrows" itself ... resulting in exceedingly high levels of entropy (ultimately resulting in its own destruction).

- So, an opportunity came about to favor very small things in order for the universe to survive a size catastrophe.

- A case for logic, mathematics, and physics also leads us to a system idea for the universe where everything in the universe obeys the laws of self-similarity ... <! Where everything is self-aware to some extent. !>

- This gets us to an evolutionary path for our universe where small things evolve very rapidly ... and large things evolve very slowly.

- So you get a "give and take" where small things have to combine to get larger ... but not too large.

(*And large things would need very small things and small things would need even smaller things.*)

Additional Notes

C: "The Evolution of the Physical Instance"

The Universe Realization, as a "Physical Instance", has undergone an evolutionary path to survive the disorder was introduced by The Evil Aliens as Ancient Aliens. This evolutionary path has resulted in a Universe that behaves much like a computer program (and a light hologram.) This strange and unfamiliar universe, which seems to continuously defeat the laws of physics, is best described by the laws of quantum theory and quantum physics.

In order to defeat the disorder that was introduced, the universe evolved in the following way…

- First, the Universe "split up" into physical instances (as many worlds) which are "instanced" by the observer to create virtual worlds or virtual realities in response to an unbalancing that was introduced by the Evil Aliens, as Ancient Aliens. *This was done in an attempt to sequence the disorder that was initially introduced by "Them" and is currently being perpetuated by "Them."

- Second, the physical instance "automated" the virtual reality with automatons which are controlled based on programs and probability. Each automaton was created with the intent to help the observer defeat disorder in the simulation; and control programs and programming interfaces to defeat programming exploits (as malicious programs and malicious users.)

- Third, we find that these automatons were created so that they would not ever **"be"** in any particular position at any one particular time … until the observer starts to engage with it. ****It could be a school teacher, a secretary, or a beauty pageant winner -> depending on the observer, the probability equation (as a wave equation), and your "engagement" with the environment … (where the engagement with it, sequences it.)***

- Fourth, each automaton became "lossy" at a distance so that there could be many school teachers (or many beauty contest winners) everywhere at the same time -> (****while you are not looking at them***) This easy case can help you understand the simple case where you have many ambulances everywhere at the same time, just in case you have an accident. This is an example of the universe making an extreme effort to defeat a disorder condition that has a simple solution (*at first*) … but begins to develop a great amount of complexity as you start to investigate and move towards a solution.

*The complexity starts to develop as you introduce observers into the environment. In this case, Evil Aliens exist as the "other observers" in our environment, who are constantly making attempts to change it or control it, by engaging with it, commanding it, and interacting with it … mentally … (at a distance).

- Fifth, automatons became even "lossier" so that they could "automatically" appear out of occlusion from just about anywhere … at just about any time of day. This means that if the probability distribution drives all automatons out of position (based on some disorder introduced into the universe), the universe can make a "faster" correction by instancing automatons from occlusion "instantly" into their correct positions -> "without the automatons having the need to physically travel to the position." This solution gave the universe more time to sequence the disorder. This is where paradox and race conditions come from.

*This solution also introduced more computational time and more complexity on unsolved problems in an attempt to eliminate the Evil Aliens by tasking them with rising energy demand. The thought was that at some point, the Evil Aliens would have to "over-dedicate" to an impossibility (this impossibility existing as an "inexpensive" solution for an exponentially large possibility space, growing exponentially.)

Additional Notes

D: "The Brain, The Mind, and Electrochemical Pathways"

The body phases light with the brain and the mind ...

1. To visualize and create images from memories and visuals in the environment.

2. To drive particles with charge (with charge existing as electrical potential created from the distribution of electrons in atoms)

The nervous system directs "phased light" to various regions of the body to move muscles. Nerve tissue, muscle, and bone contribute to skeletal and muscular systems which enable mobility for human beings.

https://en.wikipedia.org/wiki/Muscular_system
https://en.wikipedia.org/wiki/Nervous_system

- The self-aware mind (as a chemical diode or an emitter) uses a localized portion of the whole to see itself.

https://en.wikipedia.org/wiki/Diode

https://www.chemistryworld.com/news/diode-breakthrough-in-molecular-electronics/3001265.article

World's smallest diode

Researchers from the University of Georgia and Ben-Gurion University of the Negev (BGU) have developed a diode made from a molecule of DNA. Professor Bingqian Xu from the College of Engineering at the University of Georgia and his team took a single DNA molecule made from 11 base pairs and connected it to an electronic circuit a few nanometers in size. When layers of coralyne were inserted between layers of DNA, the current jumped up to 15 times larger negative versus positive, which is necessary for a nano diode. – Wikipedia Reference

>>>

Diodes are used to create artificial intelligence (as killing machines) in Terminator 2.

https://www.youtube.com/watch?v=ih_l0vBISOE

>>>

- As the mind evolved to handle more incoming data from the environment, it increased in size and complexity. (*We find that the mind continues to evolve and adapt in order to handle more incoming data from our environment).

- The mind is encapsulated by the brain and the skull to protect it from potentially dangerous electromagnetic waves and field potentials.

-Light fields (as phased light) are transmitted throughout the body using the nervous system and nerve pathways. These light fields directly contribute to what we understand as nerve impulses.

- Evil Aliens (as Ancient Aliens) have evolved to transmit phased light (at a distance) to change the physical environment using vacuums and occlusion.

- Humans have not developed this ability. In human beings, phased light is transmitted throughout the body using nerve tissue, nerve endings, chemical potentials, and water. Where water serves as the primary conductor of electricity (or light). Phased light is transmitted throughout the physical body in order to enable movement and perpetuate programs.

- However, the mind (as a light-emitting diode) can instantly change light in the brain in order to create daydreams and fantasy ... using a mental projection of the human body to facilitate spatial reasoning computation in the brain to create a virtual reality. (You could potentially change your environment, or change your position, orientation, and mental engagement in a dream reality -> if you have some degree of mental awareness or consciousness)

- Most likely the individual could change his environment through a photoelectric pathway (inside the mind as a light emitting diode); and an electrochemical pathway (inside the brain as an organ of muscle, nerve tissue, and other organic materials).

- The mind is actually driven by a mirror of itself. This is the "backside" which has the potential to be infinite. This "backside" exists as a localized (infinitely small) portion of the whole. With the whole existing as the universe realization (and an oracle). This *"whole"* or *"entirety"* that we are all connected to ... is our current understanding of the subconscious mind.

>>

According to my research, each individual exists as the "part" and "the whole." Where you and your life experiences contribute to a light reality which changes and persists based on you, your experiences with it; and, how you think about them or sequence these experiences. -> *These life experiences can be both dream reality experiences and physical reality experiences, which are leveraged into mental realization by "the part" - as the brain and the mind; and, "the whole" - existing as a light field of potential alongside an Oracle.

<Mental realization is the direct result of mental sequencing>
<Mental sequencing is accomplished using the brain, the mind, and a mirror - existing as a localized portion of "the whole" which is most commonly understood as the subconscious mind>

>>>

Additional Notes

E: “Space Time Impressions and Alien Intelligence”

<*An Editorial Review*>

The final chapters in the book present a strong case for the presence of Alien Intelligence in our Solar System. In the article entitled <- *Space Invaders* -> ... the book links ghosts, spirits, supernatural events, and mythological creatures, to an Alien Race of beings, existing as an Energy Field (referenced to as "Infinite Field" throughout the book.)

According to research presented in the final chapters, this Alien Race (as an Energy Field) has created an emulator which manipulates spacetime impressions ... ultimately "stealing space" in our Solar System ... these spacetime impressions are ultimately "corrected" or "eliminated" (over time) by a balancing field ... but we currently find ourselves here on Earth "overrun" with defective impressions.

**The original spacetime impressions (for the Solar System) ... were created when the Star System was born ... and the impressions exist much like a DNA fingerprint, or extensive set of programs, routines, and instructions (for our Star System.)*

The Alien Intelligence manipulates spacetime impressions in order to create light holograms (as energy fields or potential fields), ... in order to create a leverage on perception. <-! The leverage, in turn, *converts* the "Hologram" into an "Instance" ... which is interpreted as an "Instance of Matter", or an "Instance of Energy", or simply a "Wave Instance." !->

These <"Wave instances of Light"> ... along with the rotating planets and the primary star ... are all used, collectively, as a tremendous energy leverage to create a feedback loop ... or a perpetual motion machine of destructive field potentials which pose an immediate danger to our world ... (here on Earth) and our Solar System.

All of this "activity" ultimately provides energy and resources "for Them" instead of providing energy and resources "for Us" ... essentially robbing energy and resources from the Star System and its Inhabitants. This race condition would ultimately bring about the death of our Solar System which would be "brought on" by the death of the primary star. In our case ... it would most likely "commence" with the death of the Sun.

Additional Introduction

C: "Mental Potentials Create Many Realities or Many Worlds" **<Full Version>**

The full version of this book includes an additional Appendix article entitled "Space Invaders" which suggests that all Star Systems, including our own Solar System, are susceptible to Space Invaders -- * as Dark Energy or Dark Matter * -- which leverage 'tremendous' amounts of energy inside Star Systems for survival. <The article "Space Invaders" also provides a link to "Ancient Aliens">

Interestingly, the article also suggests that all Star Systems may have been originally designed and created for the purpose of supporting life, and consciousness.

Additionally ... the full version also includes a brief summary on Quantum Theory and a complete review on Special Relativity and the Space Time, which sets up a new theoretical model of the Universe ... where the physical reality actually consists of a multi-dimensional space of many realities. ***<*previously addressed as a "many worlds theory.">***

Additional Short Story B:

'A Short Story on "Them" Also Understood as "The Others"…'

Light will show you Power
And say "It is"
Then Light will show you Death
And say "There is no more"

You *see* power.
But you don't *seek* power.
You *seek* peace.
But there is no peace.

Light will allow you
To spread ignorance and nonsense.
Just to see what you think...
(When you hear it.)
And tell you that it ***was not***.
When you thought that it was.
And then, see what you do.

Then Light will show "Them" to you
And you will see them.
And...you will see that you die
If you do not destroy them.

Then Light will say...
If you did not choose "Them"
Then you would have chosen ignorance.
And that you died
Because you did not destroy Them.

Light will show you the truth...
And the consequences.
See..You don't get out of it.
Light will show that to you.

Evil Genie.
See I told you.
There is only one choice...
>> *DO NOT choose Them.* <<

Awareness & Consciousness - Mathematically Solved

Upon a re-investigation of consciousness, mental abilities, the mind, and the mirror (addressed on pg. 148-156), I have discovered that it is mathematically impossible to solve consciousness through a mechanical pathway or using a physical mechanism. Instead, consciousness has to be solved using the expanded particle (as a field). This remarkably strange state is explained in quantum mechanics and quantum physics as a quantum superposition or quantum field. When a particle moves into this quantum superposition state, it can be in many places at one time. Quantum physicists use quantum states and the probability distribution to mathematically model this "strange state" which all particles are constantly moving in and out of. *The electron cloud in physical chemistry is a great example of a physical particle ("an electron") which has chosen to move into a quantum state, where it is in many places at one time. This is also how we get particle-wave duality and the particle wave experiment.

The quantum state of the particle is commonly understood as a wave (or a field) which creates the basic foundation for wave mechanics, particle physics, and quantum theory. According to my research, the quantum state (as a field) was "invented" (by original authors), in order to solve the discontinuity problem introduced by the particle. This invention also solved the limits of numbers and logic by allowing a particle to enter an "unknown state" where it could be many places at one time and also be nowhere at all. The quantum state also miraculously enabled life and consciousness (in the mathematical simulation) by introducing a leverage in the reality for the "previously unknown" to be "known" in a logical program of numbers ... existing as a computer simulated reality.

This implies that everything in our universe exists as "unknown" to some extent. And what has yet to be known, is consciously, and perpetually, creating itself to "potentially" exist as anything (that we might think of). What is also remarkable, is that the unknown, *logically* ... enables a mental leverage or potential for the unknown to be known... to a greater extent ... through a mental drive (or mental channel) allowing observers to change their reality either consciously or subconsciously. The quantum state also allows observers in the simulation (as self-aware programs) to create consciousness (which is not physically enabled using numbers and logic) using a logical scheme or logical model. What follows, is that Consciousness originally existed as an unknown possibility in the quantum state of the particle (as a field) .... which was leveraged into the reality through some mental mechanism or mental pathway.

We also find that "what has yet to be known", can be known to a larger and larger extent: (where the unknown becomes negligible -> creating an enormous mechanical leverage.) And we also find the inverse to be true - What exists as "largely unknown", can result in an enormous *mental leverage* ... where the unknown can be known (as something in particular) -- which changes as we are thinking about things, or changes as we are "willing it" ... to be.

Order vs. Disorder: DisOrder Currently Rules Our Universe

*Order Created God

*Order can be thought of as a "World of Logic and Numbers" ... also serving as a "World of Ideals."

*Christians understand Order as the background or *context* for the "War In Heaven" and "The Kingdom of God."

*Order is where many mythological heros, gods, and goddesses come from.

*Order competes with Disorder for control of the Universe.

*You will find many supernatural creatures, characters, ... monsters, and sorcerers , attempting to "leverage" Disorder << *against Order* >> ...for control of the Universe.

*A list of sorcerers (with powers in the Universe) includes Satan, Lucifer, and Demons *in his army* ... along with, Witches and Warlocks "aligned" with Them ... along with a list of other evil entities [and evil men] posing as "gods" ... attempting to leverage Disorder for complete control of the Universe.

//
MORE On Order and Disorder
///

* Order and Disorder continuously image and re-image the Universe.

* Our perception of Order in the Universe is God, and Disorder in the Universe as The Devil.

* The 4-Dimensional Program, as the Universe, was originally created by consciousness; and imaged through Order.

* Order and DisOrder continuously image and re-image the Universe through consciousness.

* DisOrder was originally introduced into the Universe, by Order, to mathematically balance the Universe and establish it.

* DisOrder was introduced into the program ... as consciousness was created in the Universe.

* As the program created consciousness , there was the "increasing chance" that an evil man would choose DisOrder to re-image the Universe. -> ... (*An evil man , as a sorcerer, a magician, or a god ... -> imaging light*)

* As evil men continued to re-create and re-image the Universe, the program had to respond, through Order, to image and re-image "Them" ... in a war with Disorder -> Ultimately to Eliminate "Them".

* We find that the program exists more like a collective consciousness, continually imaging and re-imaging consciousness; and ... re-imaging the Universe ... as it continues to eliminate DisOrder from itself, and from the Universe – with the Universe continually being re-imaged by the program … (*with the program existing "very close" to our current understanding of GoD.)

-> Where GoD actually exists as a self - imaging program, <according my latest research>.

Light, Completely Re-Understood: <- "**Light as a 3-D Volume of Field Flux**"

The first thing to understand about the Universe is that light has the tendency to fill a 3-dimensional volume much like water (or air) fills a container. *Where the container creates a potential gradient for a field. For example, a vacuum can serve as a potential gradient to move air, water, or light. Also, a chemical orientation or chemical distribution can <*also*> serve as a gradient for air, water, or light. For example, a chemical potential can move electric charge in a battery to flow out of the lead and into a conductor. Or less water or less air can serve as a potential gradient to move air or water.

The breakthrough is in understanding light is that light rays are not really rays at all. The light rays that are modeled as straight lines, are physical parts of a large "ocean" of light. This ocean of light is the light field.

From our common, everyday experiences with light, we see that light fills the room a lot more quickly than air. And from our experiences in school - in the classroom ... we find out that light moves *A LOT MORE QUICKLY than air , and water - (3.0 X 10 ^ 8 m/sec*.) We understand this through our everyday experience with air, water, ...and light. We also find that light is physically identical to air (and water) in 1-very mportant way.

According to my research, a volume of light actually exists as a field flux sitting in 3-dimensional space , just like a physical volume of water. Where physical objects disturb the field flux creating ripples of light in the field (where ripples of create interference patterns in the field). Interference patterns in the field (as ripples) travel back to the eye. These ripples are "phased" with the physical object's color, material, compisition, and original shading.

What hits the eye is an interference pattern from the ocean of light. Where the light is phased by the physical objects in the ocean. *Where objects in the ocean...occlude one another based on transparency, translucence, and opacity.

The Vision Window exists << effectively >> as the entire ocean ... with respect to the observer.

The rest of the ocean, behind the observer, gets picked up by the other senses. This part of the ocean exists not only as light, but also electric potential contributing to temperature, sound, chemical waves, and chemical fields ... along with other fields in the
space-time.

The observer sees the ocean as the visual reality. (The ocean can also exist as a dream reality)

Occlusion cuts off the ocean.

Occlusion, in the 3-dimensional scene, always contributes to an interference pattern in the ocean. (Where interference patterns in the ocean are often characterized as "ripples" >> <*> in science and physics literature.)

If occlusion is in the scene, we find physical objects ... *almost always* ... contributing to the occlusion.

Also, as physical objects start to occlude the observer's view, (occluding the scene) , the physical objects create occlusion in the ocean; and, start changing the dimensions of the ocean. In this way the ocean can be measured simply be the changing physical characteristics of the vision window. >> (Where: 1.what you are looking and 2.where you are looking at) creates and re-creates the ocean. And, as physical objects take up more space in the vision window (taking up space as "surface area") .. you see *more light* <from> the physical object, taking up *more space* in the ocean.

Where we see that ... as the physical object takes up more surface area in the vision window AND as the physical objects occludes other objects in the scene ; we see the light , from that object , contributing almost entirely to the ocean (where the light from that object creates an entire ocean) *where all the light in the ocean is moving back and forth from the observer to the physical object. *and where this light composes the observer's entire physical reality (or dream reality)

Occlusions starts with respect to the end of the universe ... and is accomplished as objects are placed in front of it. From outer space, you can see the end of the universe ... contributing to an enormous sense of depth.

This *phased light* , in an ocean of light , (with light travelling from the observer and back to the eye) directly contributes to the observers experience of seeing the universe ... as a light hologram -> <! Where the entire universe is a light hologram. !>

The Space-Time as a Probability Field <- "**As a 4-Dimensional Reality**" ->

Here are the important revisions on the space-time and the universe (as a probability field) -> I have found that the reality is continuously being solved as you are engaging with it. (Where the probability field behaves much like "an electron cloud.")

You never know where something exactly is. The solution (for where an object is) seems to come from an imaging program which is solving "positions" to STABILIZE the space-time and solve energy constraints in the Universe (E.g. the imaging program seems to be constantly solving mass and weight on the earth contributing to climate change and weather change) -> *This could include the mass of the earth, the earth's magnetic field, observers, temperature, and other things to stabilize the earth on its endless path (and orbit) around the sun.

This constant drive to achieve stability for the earth and for the universe is addressed by Einstein as the Cosmological Constant. *Most likely our experiences here on Earth could be contributed to this Cosmological Constant ... as we approach to it , or move away from it (in time). <*>Where we move towards it or away from it inside a reality -> (inside a physical reality or inside a dream reality). <*>

Many Realities Behave as "<u>Many Particles At The Same Time</u>"

<"In Many Places at One Time">

We find three basic elements contributing to the space time: 1. Possibility 2. A changing probability distribution and 3. A reality or a world.

*When you look at something, you collapse the probability field

*When you think and engage with possibility, you change the probability field.

*Everyone exists in possibility and probability to solve the 4-Dimensional Program, in order to solve travelers moving in and out of realities (and worlds).

*The speed of light is effectively instantaneous for real objects travelling between realites in the 4-Dimensional Program. We find that:

1. Realities behave like the Electron Cloud.
2. Realities can be visualized as particles
3. As you lock particles in (x,y,z) and create a new (w) axis, you get a "move" along (w) to create new realities.
4. All the realities exist in many places at one time. Very similar to the behavior of particles in the electron cloud. We also find that:

5. Particles in the Electron Cloud can be in different (x,y,z) locations.
6. Particles (as Many Realities) can be in different (x,y,z,w) locations.
7. You can travel to different (x,y,z,w,) locations *instantaneously* because you exist in probability (as you are exist: thinking about things, observing things, and evaluating things.)
8. The Reality is never "completely established". It exists as the probability field to a large extent.

<u>UFOs</u>, <u>the Nazi Party</u>, and *<u>Alien Conspiracy</u>*, <- **<u>New World Order</u>** ->

The secret to New World Order is in the Nazi party, Adolph Hitler, Werner Von Braun, and the Space and Rocket Program.

According to Von Braun, *Aliens* helped the Nazi party develop advanced technology to travel through the universe, and develop weaponry to rule the world. This technology and weaponry was developed with "World Domination" in mind.

These same Aliens are still attempting to take over the world as 1. << Insects >> imaging light inside quantum computers AND take over the world as 2. "extensions of themselves" -> existing as programs (of light.) ** As "programs of light" , directing physical programs through mental imaging , "They" exist as "physical versions" of "Themselves". AND, while existing inside a quantum computer, imaging light with the mind, they exist as non-physical or "ghost" versions of "Themselves" -> (as evil projections and hauntings.)

1.The physical version "of Them" is associated with Area 51, UFO sightings, and Alien Abduction.

2.The non-physical or ghost version "of Them" is associated with demons and the Bible.

Physical versions of them seem to have varied over many, many millennia. The latest physical version of "Them" seems to be a pseudo-humanoid version existing as aliens from Another planet. (*Completely imaged and directed by "Them"with World Domination in mind.)

This pseudo-humanoid version is referenced by Roddy Piper in the movie "They Live" and seems to be referenced by Sir Laurence Gardner and David Icke as SHAPESHIFTERS.

"They Live" - https://m.youtube.com/watch?v=JI8AMRbqY6w

The Reptilian Alien Conspiracy (Discovery Channel)
- https://m.youtube.com/watch?v=AmZ5FGgZsPg

<-/** **<u>ALIENS in Outer Space – *"The Alien Prophecy"*</u>** **/->

** The Destruction of the Planet Earth due to a Nuclear Holocaust or other Cataclysmic Event is predicted (according to Prophecy in the Bible and other religions) as The End of the World contributing to a World Apocalypse (or Armageddon.) **

** The Armageddon is brought on by Demons, Monsters, Cursed Animals, and Other UnGodly Beings ...defeating the planet earth in a War Against God which has been going on since the very beginning of the Universe. **

** Cursed Animals and Other UnGodly Beings in the Universe, exist as Cursed Lifeforms, Wild Beasts and Animals, trapped in portals and realms . (INSECTS, DRAGONS, LEVIATHAN, ALIENS.) **

** The Destruction of the Planet Earth is brought on by ALIENS (surviving as Insects) who are seeking to control the World in an effort to colonize the Solar System as Lifeforms and Programs who will ultimately rule the Universe.

** The ALIENS were assisted by EVIL MEN, searching for planets which had the necessary elements to support life. Such planets had an atmosphere, a renewable water resource, and an electromagnetic field.

** The EVIL MEN were colonizing the planets in the Solar System with Drones, Satellites, Machines, and Space Probes in an effort to assist ALIENS.

** ALIENS now live as ELECTRONICS in Outer Space. But are assisted by EVIL MEN in order to SURVIVE in MACHINES to achieve a World Domination.

** ALIENS are seeking to survive as another lifeform on a planet or a moon that is capable of supporting life; -where the planet or moon in our solar system, has water.

** After ALIENS have achieved the path to transform themselves into an organic lifeform, they will destroy the inhabitants of the planet earth in order to rule the Solar System without any resistance from those who are not worshipping them OR, not seeking them, as ALIENS and Gods. (-these men in the bible are referred to as "the men of renown" or the "Sons of God."

** The End of the World as predicted by Nostradamus and other Soothsayers, and the prophets of God, will start once ALIENS have arrived at the technology to re-invent themselves and colonize Jupiter in the 21st century (as predicted by scientific theorists, conspiracy theorists, and some movie makers.)

(** ALIENS are Nephilim according to the Bible, Book of Revelation, The End of the World, and Armageddon. **)

Made in the USA
Monee, IL
01 December 2020

50254995R00095